Lazarus in the Multiple

Awakening to the Era of Complexity

Lazarus in the Multiple

Awakening to the Era of Complexity

Camaren Peter

Winchester, UK
Washington, USA

First published by Zero Books, 2016
Zero Books is an imprint of John Hunt Publishing Ltd., Laurel House, Station Approach,
Alresford, Hants, SO24 9JH, UK
office1@jhpbooks.net
www.johnhuntpublishing.com
www.zero-books.net

For distributor details and how to order please visit the 'Ordering' section on our website.

Text copyright: Camaren Peter 2015

ISBN: 978 1 78535 108 2
Library of Congress Control Number: 2015941041

A CIP catalogue record for this book is available from the British Library.

Design: Lee Nash

Printed and bound by CPI Group (UK) Ltd, Croydon, CR0 4YY, UK

We operate a distinctive and ethical publishing philosophy in all
areas of our business, from our global network of authors to
production and worldwide distribution.

CONTENTS

For Vanya and David

On the Purpose of this Book:
Meditations, Reflections and Imperfection

Now that I have finished examining the judgements of others, I return to the same questions of God and the human mind and to the beginnings of the whole of the First Philosophy, but without waiting for popular approval or wide readership. Indeed, I would not encourage anyone to read these pages unless they are willing and able to meditate with me seriously and to detach their minds from the senses and simultaneously from all prejudices, and I know that there are few such readers. As for those who do not bother to understand the order and interconnection of my arguments but try to snipe at individual sentences, as they usually do, they will derive little benefit from reading this book. They may find an opportunity to cavil in many places, but they will not easily raise any objection that is significant or deserves a response.
—*René Descartes* (Meditations, 1641)

This book consists of a series of meditations and reflections which, taken together, constitute a philosophical discussion in the form of an assemblage. As such, the purpose of this book is not to 'formulate' or 'construct' a perfect theory, but rather to open up new avenues of discussion on long-held philosophical beliefs that, in some quarters, have come to be considered axiomatic in nature[1]. And for a long time, it has burdened me that such meditations and reflections may be misconstrued, taken as presented as fact, when their purpose is quite the opposite. They are presented here to be meditated and reflected upon, and not regurgitated as special knowledge of any kind.

And in order to ensure that the content of this book is taken up in meditation, I have made every effort to write these meditations and reflections so that they *resonate* with the reader first, even before the reader comes to a full understanding of the concepts that are introduced and discussed in this book. It is in this initial resonance that I have put my faith, so to speak, to serve as a catalyst for deeper enquiry on the part of the reader. This is not a prescriptive work; it is a work designed to open up trajectories for new understandings and insights to emerge. It is intended to resonate with the reader, so that the reader is disturbed or displaced, and that the reader can thereby be led to formulate questions that they themselves pursue.

And to be sure, it cannot be guaranteed that the content of this book will resonate with all those who read it. Such an objective is patently impossible to achieve, even if it is an admirable one. All that can be guaranteed is that the reader will surely encounter new knowledge and new understanding by engaging with the content of this book if he or she is able to suspend their preconceptions as they would for a song, or a poem, and dive into it with curiosity, interest, sense and emotion.

No perfect knowledge is presented here. It is imperfect as we are, as we observe ourselves in an imaginary mirror; which reflects us at the world, and the world back at us. It makes no pretence of being an academic piece of work, nor does it attempt to provide clear answers. It delivers new knowledge, but more importantly; opens up pathways to new knowledge. All knowledge presented in this book is merely put forward for this purpose, and this purpose alone.

And so for readers who are concerned with grand theories, they might find little use in this book. The content of this book is for those who suspect, deep within themselves, that there is a fundamental irreducibility to reality that we can never dispel, despite our best efforts. It is for those whose intuition indicates that reality is lost, reduced, by being abstracted. This, at least, will ensure that there exists cause for resonance to occur within the reader; that there is fertile ground—so to speak—for these seeds to germinate, grow and flourish. And as each tree and plant grows uniquely, it is to be expected that the content of this book will be taken up in different ways. I do not seek to limit the extent of the exploration of the concepts and ideas communicated in this book. Rather, I seek to set them free upon the world, and into other minds, so that they may take whatever form suits them and the purpose to which they are deployed.

We lived many lives in those whirling campaigns, never sparing ourselves: yet when we achieved and the new world dawned, the old men came out again and took our victory to re-make in the likeness of the former world they knew. Youth could win, but had not learned to keep: and was pitiably weak against age. We stammered that we had worked for a new heaven and a new earth. And they thanked us kindly and made their peace.

—*T.E. Lawrence* (Seven Pillars of Wisdom, *1936)*

Preface to Part I:
A Homage to the Muddlers

So I finally turned 40 years old last week. I feel as though some kind of equilibrium has been reached, and that it demands my attention. I have now been free for 20 years. I am finally free for as long as I lived under Apartheid. Something about that seems magical to me, and begs reflection. What is left of my wounds? What is left of my memories? Perhaps more than I would care to admit readily. Now, more than ever, I am undecidable; a duality. There is no binary to the oppressive and the free periods of my life. It is all a mix.

And this mix is something I have become increasingly aware of over time, but which I am now acutely aware of. It is as though there is a constant noise in me, permeating my heart and mind without pause, lulling only when there is clarity, a signal that I can focus on to escape the noise. I have tried to escape it, but no amount of alcohol, drugs or food can dampen it out completely. It is now a part of me. I doubt I will ever escape it; perhaps it persists beyond death. At its core, it seems in every way like the cosmic background radiation of the universe; a product of the beginning, and at the same time a dynamic imprint that connects me to everything.

Yet it is also without us, and contingent on the dynamics of environment. Now, more than ever, do I understand the multiple; *as it is within us, so is it without us.* We have more than just a physical membership to the multiple. We make the world multiple, because we are multiple within. Penetrating the noise like a deep-sea diver; or dampening the noise through employing filters and lenses to sift through it, does not undo the multiple. It is resolutely with us precisely because it resides within us and frames everything we observe and experience.

And thus, my enslavement and my freedom intertwine. They

cannot separate from each other, rendering me undecidable, and none more so than in my current disposition; half and half. Two pasts vying for me with equal conviction, and constituting me as multiple. Why this obscurity you may ask? Why this obfuscation? Well, that is the purpose of this book. It is to expose the multiple within and without us, and particularly its continuity, its inseparability. The undecidable *is* explicable, even if it remains undecidable, and there is power, merit and value in this casting of it. Lazarus—that is, the 'everyman' of modernity who is caught between life and death, sleep and awakening—is the agent through which I explore and reveal this duality, and South Africa serves as the temporal venue for this exploration. Its spatiality is not emphasised, but rather its temporality, as this allows for greater abstraction of this exploration, and its application to a broader, more global set of challenges that we face as humanity. This is not to suggest that spatiality is irrelevant to the multiple, but rather, to limit this inquiry and focus its relevance more keenly on the historical, the present and the future.

And while Lazarus is the agent of exploration, the 'multiple' serves as the 'filter' or 'lens' of analysis. I have chosen it because only through the multiple can we approach complexity and assemblage without drastically affecting it, transforming it from what it *is*, into a reduced form that enables only partial analysis. And thus, ironically, the apparent obscurity and obfuscation of this approach, in reality, attempts to preserve, as far as possible, the mix that resists reduction, and indeed loses itself in reduction, becoming abstraction instead when reduced to its elements.

Part I

Meditations on Time, Memory and Existence

Chapter 1

Multiplicity: Many Voices: Past & Present

Never have I felt as strongly as today that I was devoid of secret dimensions, limited to my body, to the airy thoughts which float up from it like bubbles. I build my memories with my present. I am rejected, abandoned in the present. I try in vain to join the past: I cannot escape from myself.
—Jean-Paul Sartre (Nausea, 1965)

No-one can reverse the past. It cannot be reclaimed, yet it is in the present, with us. It is distributed throughout the multiplicitous sea that constitutes the present—in us, between us, and outside of us—but it takes no absolute form. Its distribution across time, space and people gives it the power to emerge in different ways, depending on the context that governs its emergence. Its traces emerge from the seas within and without us, elusively roping into and out of the interactions we have. In this way it is constantly being rewritten and will always continue to be.

We call this history, but by giving it a name we trap it in a bottle in which it is unable to breathe, and preserve a dead, static version of it, and with it a dead static version of ourselves. The question: "whose history, whose past?" allows us to destabilise the lifeless presentation we call 'history' because it forces us to confront the distributed nature of the past as it is understood and experienced in the present. The past is not history, and history is not the past. The past is still with us, emerging in different ways as we journey through the present, precisely because it is distributed in all of us.

While history can be rewritten with the vagaries of power and politics, the past is elusive. It retains agency by living within us.

It is individual and collective, personal and shared. It is living and breathing with us. It is part of our interactivity and non-linearity; we cannot be neatly arranged into hierarchies and neither can the past. No hierarchy can hold the past hostage because the past is more than what can be written or recorded. To engage with the past as it exists among us is to engage deeply; to immerse oneself in the sea of the known, the unknown and the yet-to-be known, and to swim in the present; wide-eyed, aware.

Even our personal, individuated pasts are, ironically, not completely known to ourselves. As we age, we become more aware of how the past walks with us; often catching us out where we least expect it. This is where the past emerges. Sometimes it can be understood in terms of the personal history we believe in, which is reinforced when patterns of behaviour that emerge in the present seem to match those that existed in the past. Here, a history—even a personal history—can be useful, but if it becomes a dominant, self-referential narrative it impinges growth, stunts life.

The past—inherent, latent, distributed—comes to life when it emerges between us, in our interactions. Here, it takes on recognisable forms, and becomes a traceable 'thing' from the past, a thing that only we can validate and breathe life into. We coax the past into existence in the spaces between us, and the more we reinforce it the more it grows in the present, coming alive like a Frankenstein beast; a zombie, an undecidable[2], both dead and alive at the same time. As it grows and strengthens between us, so it consolidates the ties that bind us to each other, and to ourselves.

A self-validated 'history' is but a snapshot of this. It is more concerned with the specificity of events and context, and less with the patterns of behaviour that self-replicate across time, ignoring spatiotemporal, personal and interpersonal boundaries to emerge as a many-headed beast; distributed heterogeneously, yet still taking on recognisable forms of behaviour. What is distributed brings about replication and self-similarity at different scales of inquiry and aggregations of analysis. The 'past' comes alive and breathes in the present. History, by virtue of its methodological foundations, is ironically 'stuck' in the past by virtue of being constituted as written, artifactual, audio or visual records. Understanding 'history' as it exists requires our absence—while engaging with the past as it exists requires our presence.

The past is, in this way, *present*. It is not in the present, yet it *is* present. Its absence is merely a sinking into the 'noise' of multiplicity, where it appears as nothing because it has joined with everything; a void of perfect noise. Its absence—or silence—is merely when it 'disappears' into distribution and lies dormant with no context to activate it. In order to know the past, we have to draw it into the present, see it in the present. Cycles of war, conflict, state-abuse, human slavery, genocide and ecological exploitation ("ecocide") continue unabated, silent only through brief moments in history. We have to acknowledge that it is here with us, so we can recognise how it self-replicates with us.

The present contains everything. It *is* the continuum of the past and the future. It gives life to both the past and the future— they can only live through the present, and by implication; through us. What do I mean; "through us?" By this I mean that the present is what we experience. As such, it is intimately dependent on observer-subject duality. What may constitute the present at any given moment may be experienced uniquely by different people. While the reality that is co-created at the collective level may—at the global level of behaviour—seem coherent, invariably at the detailed levels of experience we tend to differ.

The multiple as such. Here's a set undefined by elements or bound-aries. Locally, it is not individuated; globally, it is not summed up. So it's neither a flock, nor a school, nor a heap, nor a swarm, nor a

herd, nor a pack. It is not an aggregate; it is not discrete. It's a bit viscous perhaps. A lake under the mist, the sea, a white plain, background noise, the murmur of a crowd, time. I have no idea, or am dimly aware, where its individual sites may be. I've no notion of its points, very little idea of its bearings. I have only the feeblest conception of its internal interactions, the lengthiness and entanglement of its connections and relations, only the vaguest idea of its environment. It invades the space or it fades out, takes a place, either gives it up or creates it, by its essentially unpredictable movement. Am I immersed in this multiple, am I, or am I not a part of it? Its edge a pseudopod takes me and leaves me. I hear the sound and I lose it, I have only fragmentary information on this multiplicity.
—*Michel Serres* (Genesis, 1982)

Likewise, the past emerges heterarchically[3], bound to the context of its creation in the present, and as it is lived out by people. As the present is translated to history, the same applies—our analysis of history occurs in the present. Even when we make use of 'artifactual' information and evidence; it is analysed, brought into life, in the present. The present knows no absolutes, only uniqueness. Multiple patterns of behaviour are distributed throughout the present at different scales and levels of description. It knows no hierarchies; it is unbound by space and time yet distributed across it. Its vast multiplicity—its fullness—subverts and obscures attempts at categorising, ordering and formulating. Its fullness creates room for elusiveness, for variability and undecidability. Fullness creating emptiness; or both co-creating each other; the dance of Natarajah—the dance of reality.

In the present, the past is expressed through many different voices; written, virtual, musical, poetic, verbal and non-verbal; all that gathers in the accumulated humdrum of everyday life. So how do we accommodate this 'noise' yet still recognise patterns as they emerge, perhaps early enough to break undesirable patterns or reinforce desirable patterns? How do we engage with the many voices of the past that still shadow us; haunt us? What surprises us, scares us, delights us, or leaves us ambivalent? We all negotiate the terrain of the past both differently and similarly. Our senses come alive to engage with different levels of experience, as do emotions. Whenever we think, talk or act we bring the past alive. The past emerges, indivisible from the present, distributed in all of us, and is primarily recognisable in our interactions.

Can the richness and multiplicity of human society be captured by notions of 'nationhood', or does nationhood as a principle restrict human society, and just as religion and Marxist Leninism have been abused, find *itself* abused as a tool and principle for organisation of society?

The danger in composing dominant historical narratives to unify the nation state is that as it rises to the 'meta-level' it robs us of the ability to recognise the past as it emerges in everyday life. Suppressed and unacknowledged it churns and bubbles up in the least expected situations. This is what makes our versions of history vulnerable (and us with it). It is what makes us a *human* society. As individuals, we journey through various notions of personal identity in a single day (or even a single interaction) with ease. This is more what makes us a human society than what defines us as a single 'thing' or another, where we exist in the binary. In the desperation for a sense of nationhood, we cannot forget what the dangers of unbridled nationalistic sentiment can bring. We are many voices, many pasts, all combined, all distinct. We are a humdrum, a multifarious many-faceted noise. We are the vapour in which rainbows appear. We are not the rainbow. All attempts to crystallise us, undo us for what we are.

In this humdrum the past becomes a shadow hidden in the low probability zone of the spectrum of everyday experience, discreetly hidden in the noise, a shadow trying to wrest itself free of the pavement of the present; emerging in non-linear bursts out of the two dimensions of the flattened pavement and into reality. Little packages from the past make their way through to the present in this way, through our voices. Crime, war, forced migration and poverty are all repetitive spirals, cascading through the ages, making our reality indivisible from the past, distributed throughout all humanity. So are human compassion, altruism, struggle, redemption, reconciliation, compromise and genius. The reconciliation that supplanted revolution in South Africa has occurred a million times before in everyday interactions between South Africans throughout the ages. The rainbow emerges and so does its shadow; reality is co-created in the dance of duality. Everything that has happened before happens again, anew. In every space we interact in, the past appears and is voiced. All that is missing is the ability to hear it, to see it, but it is felt. It rises up within our chests and strangles us for a moment, whence we turn away, release it, and pass it by again; but it is resolutely here, shadowing us through the present. And so we need to go beyond merely showing and explaining our history to each other, towards exploring how we live with our past as it manifests in the present.

Chapter 2

On Forgiveness and the Future

The future is also irreducible because it is multiple. In it lies uncertainty, surprise and impossibility. Forgiveness in its ultimate realisation, as accounted for by Derrida (2001), is contingent on "forgiving the impossible". It is not mere "reconciliation" or *pardon* that is given for the sake of maintaining stable, if conditional, relations in society; where the subject and object of forgiveness are relegated to individual accountability for both forgiving, and being the object of forgiveness in society, respectively. True forgiveness strives for unconditional forgiveness, even if it never quite achieves it.

> *So I asked Derrida; what's the difference between an undecidable and the impossible? He thought for a bit ... and replied, "Well that's an undecidable!"*
> —*John Comaroff (2007, private communication, anecdotal)*

The question remains that: "once the impossible becomes possible, then what is left of impossibility?" Impossibility dissolves and possibility emerges into the present, experienced in multiple ways by the many occupants of reality. The future becomes the present and the realm of the possible emerges; one future has not emerged but many, and for many. By contrast, what is *probable* is a projection of the past as it is drawn from the past. Therefore, in navigating the future we must necessarily consider the impossible as it is this from which the possible, and the future, emerges.

> *The problem is that we are all blind, all dependent on preordained representations, on what we think we'll see. Most of the time, that*

is how it is. We don't experience the world. We experience our expectation of the world. That expecting is really, really complicated.
—*Hustveldt, 2008 (in Van Marle, 2009)*

Is it possible that forgiveness can emerge, as a surprise, from the future; spread out in different moments and interactions across the social spectrum in different scenes of life because of the power of the individual to appropriate the past and the future in their own, unique way? We have to believe it is possible, otherwise we can never truly be free of the ghosts of the past, and our goal of liberation will remain lost forever. If our quest is freedom, we need to believe that we can put our ghosts to rest. The probable cannot restrict us in this vision because it only consists of the past, and history; it contains nothing of the future, and indeed, then what is the point of vision itself? What 'we think' is possible is linked to our expectations, and can range from the probable to the positively outlandish (for example, we often prefer the security of superstition to the uncertainty of chance), so how can anything be unforgiveable when the continuum between the impossible and the probable remains so porous in our minds, and in its emergence from the future into the present? Impossibility is part of both our internal and external realities in everyday life; it is not out of reach; it just cannot be reached by design alone. It has to emerge. It cannot be controlled, or it loses its real power; the power to liberate.

So in forgiving, do we invest in the 'who' or the 'what' (Derrida, 2001)? Do we forgive, reconcile or even forget? How do we negotiate past and present in the future? Do we forgive the 'what' or the 'who' for the transgressions or violations that have been imposed on us?

The past is important for what may have been, or what may be, but in order to conceptualise a future we have to ask the question *"who* may be?", and "how?", and "why?" We cannot conceptualise a future independent of ourselves. This both limits and empowers us; in a converse duality of being able to shape the future on one hand, but being powerless to control it on another. Our capacity to act, but incapacity to control or anticipate, is the condition of the 'who'. Therefore, it is through the 'who' that we must navigate the future.

When we truly forgive we invest in the 'who'; that is, we invest in ourselves. That is the secret. To some extent it is a selfish act, or an empowering act. To truly forgive another, one first needs to forgive oneself; that is, forgiveness emerges from self-reflection within the 'victim', which is then projected upon the external 'enemy', or object of forgiveness. So the other is always within us, and must be negotiated within us for true forgiveness and resolution to occur.

To forgive is to abandon the self-narrative of victimhood. Does holding the discourse or narrative in the binary framework of victim and perpetrator in reality prevent 'real' forgiveness, or work against it? For example, does constraining of the narrative of the Holocaust to that of a perpetrator-victim drama also prevent real forgiveness from occurring, and does this conversely allow Israel to maintain a policy of suspicion, aggression and separation from Palestine, and for Palestinian organisations to maintain their policy of suspicion and aggression in response? The Holocaust was a crime against the individual justified as a crime against the multiple. Is not responding in kind—i.e. caricaturing the perpetrator as the multiple—committing the same error, at the root, which lies at the foundation of all genocides? In short, perpetuating an 'us' and 'them' binary is what continues the cycles of exploitation, oppression and hate, and prevents them from being broken. The perpetual requirement for a victim and perpetrator (or conversely, forgiver and object of forgiveness, respectively) means that the narrative is framed within these terms, and cannot operate outside of them, preventing true freedom, pluralism and diversity.

Ah, but one may ask, "surely no-one can ask forgiveness for the Holocaust, or Apartheid, or slavery?" ...But what if no-one was offering, or asking? What if victim and perpetrator can, in the 'who', find forgiveness and liberation? If the 'who' is indivisible from subject and object; where all objects become subjects, wafting in-between moments of attachment and detachment—the individual 'who', which when aggregated into the multiple becomes indistinct from a number—then is this the limit of our understanding of the who within and in relation to the multiple? Does the multiple obscure attempts to reach the who, and must the multiple hence be abandoned in the quest for forgiveness? Indeed, is it a race for who will be the first to break the pact between victim and perpetrator, or will mutual indifference triumph?

How can one be the object or subject of forgiveness if it is extended to the multiple? One and the multiple are not the same. The whole becomes an object or subject itself. Indeed, can the multiple be forgiven? An individual—a subject or object—can be forgiven, but not a multiple. Can we, as ordinary people, forgive the *whole* of humankind's evil because we arrive at the understanding that all that is evil is part of our own condition, or rather, a reflection or projection of our own condition?

Except in the absolute, messianic sense, the ability to forgive all mankind is indeed rare; but in all cases of forgiveness of the impossible, if not the absolute, it comes about through the transformation of the self. True forgiveness in the case of 'forgiving the impossible' requires a release or 'letting go' of memory (but not forgetting); a release that can only occur in the forgiver alone. It is not a transaction or a negotiation, that is, it is not *conditional*. It is achieving this 'unconditionality' that seems impossible, precisely because it cannot be addressed through the transactional or relational. But after this release, the forgiver is liberated, maybe not entirely, in the messianic sense, but nonetheless newly liberated from the torture they relive in the memory of suffering.

It is a new understanding.

As Steve Biko identified rather clearly: liberation emerges within the oppressed at an individual level, in a mode of self-reflection (Biko, 1987). 'Decolonising the mind' oneself is the objective; not changing the world but living with it as a trans-formed, liberated individual within the multiple. This is how liberation and forgiveness are linked. To be free, one starts by forgiving oneself. A narrative can only be discarded from within. When forgiveness is purely political, aimed at the multiple, it is only of use to politicians, states and economies. When it is personal, it brings about freedom because it allows for new narratives to be written. It allows for participation in the future.

If the objective of the state or government is control of society, then true forgiveness works against their intentions. This is because our current political, social and development discourse relies on memory. True forgiveness would create new futures, and with it, new memories and new histories. Herein lies the ironic paradox hidden in the pragmatic solution of "forgiving but not forgetting" as outlined by Levy & Sznaider (2005) for the purposes of maintaining social stability; in that true forgiveness does not occur at all, justice is not seen to be done, and rhetoric and platitudes govern the discourse moving forward. These all, in the end, lead to the dominance of modes of alternate suppression and festering of resentment, and the resulting condition can paradoxically prove more explosive for society, state and government alike, contrary to its original intentions to bring about peace and understanding, and to move beyond these destructive societal modes. Moreover, the original aim of control—i.e. to maintain stability—can easily be perverted, and it can be argued that the prerogative of control can be held respon-sible for the failure in total. To put it simply: controlling the process of forgiveness and restitution and institutionalising it can remove the process from society, whose individuals are then unable to access legitimate processes of forgiveness and restitution in their everyday lives, and on their own terms.

This is true for both so-called victims and perpetrators alike. In South Africa, the ex-perpetrator, or so called 'settler', is relegated to a one-dimensional response to the prevailing perpe- trator-victim binarity in the discourse, and hails back with claims of 'reverse discrimination'. A justifiable response in the circum- stances, but contrite in its shock at being relegated to 'the other' so quickly in the post-liberation discourse. Accordingly, it is a controlled response to an attack of control. It is predictable, and that is what is disturbing about this whole picture.

The main obstacle to diversifying the discourse on forgiveness and restitution is that it works against the intentions of the state and government. Carefully managed narratives would have to break down, and make room for others, changing the way insti- tutions and individuals understand their relationship to the complex issue of navigating a future based on an 'unforgiveable' past. And it is true that this past is lived out in many different interactions in the present day, yet its expression in society is limited by the prevailing discourse. True 'freedom' in this regard is unallowable because it threatens the existing power relation- ships that are enabled by the prevailing discourse. Where does Black Economic Empowerment (BEE) go? Where does the Reconstruction and Development Programme (RDP) go? Where does land restitution go? Where do the unions go? Where does the South African Communist Party (SACP) go? Indeed, where do the heroes and heroines of the struggle go if everybody else's story can be told freely; that of ordinary, everyday injustices that stripped people of their rights and dignity under Apartheid and continue to do so today, of perpetrators whose complex narra- tives remain unexplored, and for a more reasoned, in-depth understanding of the realities of villainy and the heroic?

At stake is how identity is allowed to form within the spaces provided by the prevailing discourse, or on its fringes. It is always on the fringes that more interesting things happen, as more goes unmonitored and undetected, or is simply ignored because it exists on the fringes. The marginalised in South Africa, in this case the majority, are relegated to the fringes of the discourse on forgiveness in South Africa, and it is for this reason that the failures of the state translate into explosions of resentment and anger. While the stated reasons are often legitimised by claims of lack of adequate service delivery or failures in service delivery (which are indeed true claims), the levels of outrage that are expressed can be found in the deeper annals of South African history. They are nothing new, and began under the Apartheid state. In the colonial and Apartheid history of South Africa the main crimes against the majority have been perpetrated through the desire for control—i.e. of people and resources—and the consequent manner in which the policy, legal and institutional arms of the state were coerced for the purpose of maintaining an unjustifiably inequitable socio-economic arrangement, of which the broad citizenry were acutely aware. Their reaction was protest and acts of social, economic and political sabotage. The Apartheid state desired control, because it needed to control dissent that was directed towards the imbalance of power that it maintained.

So why, might one ask, do the now free citizens attack a government run by their heroic liberators, as if they were one and the same government of old? Why the same response? Because the same condition exists today. New actors, old actions.

Due to the vast socio-economic differences that exist between the majority and the middle classes and elites, control remains the main paradigm for governance. What hasn't changed in the reality of everyday life of the majority in relation to the state and government is that the 'new' state and government also govern with the primary objective of control, which necessitates limiting freedom in order to legitimise power positions that dominate the prevailing discourse on black economic empowerment, socio-economic 'upliftment' and development. Ironically, this may turn out to be its undoing.

Yet we must return to the central question if any kind of progress is to be made; that is, 'how to forgive the impossible?' Indeed, it may itself appear to be an impossibility.

In order to access the impossible we must engage on multiple levels so that we can disrupt our understandings and allow space for new understandings to emerge from a basis of consensus within diversity, i.e. a consensus with multiple loci of convergence between diverse participants and actors. Engaging on multiple levels necessitates sharing, disrupting and reconstructing narratives of the past, present and future in terms of the possible and the impossible, as much, if not more, than in terms of the probable. For plurality, no single narrative—except that reached through consensus—can be allowed to rise to authority. So how do we achieve this non-hierarchical, complex, adaptive hierarchy—i.e. a heterarchy—of narratives, explications, hypotheses and postulates? Fundamental to this is speech and dialogue. It is where we engage with complexity directly. We are complicit in allowing ourselves to be held victim to uncertainty, rather than seeking to exploit it. Pragmatism reigns where decisions have to be made in complex environments.

So what of vision then, and participation, and how do we ensure participation in both visioning and action so that pragmatism, the banal, mediocre and mundane are thwarted? To ensure participation in the future we must entertain the impossible. Otherwise, we close ourselves off from the future, and it occurs despite us, and not with us. We remain cocooned until the web of illusion is shattered and reality, so to speak, bites. Metamorphosis or death is the consequence; the fabric of illusion is ruptured either too early or too late for the convenience of its occupant. Adaptation and transformation are necessities for negotiating the future precisely because not everything can be controlled, and neither can forgiveness. Forgiveness must be allowed to emerge from amidst the din of prescribed gestures of reconciliation, to take its own shape in society.

We must allow ourselves to think, speak and act unpredictably in certain situations, as unpredictable trajectories have the potential to unlock deadlock, and bring about release, i.e. to *steal* our forgiveness from the future by daring to think and act differently. This is not an argument for simplistic notions of anarchy, but an argument for holding open the space for subversion, innovation and change where notions of forgiveness are concerned. National reconciliation efforts can, ironically, work against real reconciliation, resolution and restitution by controlling the narrative too tightly. True freedom can emerge only from diversity and pluralism, as this allows us to engage with multiplicity and live in a society of choice. True forgiveness can emerge only from this freedom within, and brings about only this kind of freedom without.

What does the future hold for us if we maintain a tired, ill-fitting narrative of ourselves as victims or perpetrators confined to living out this drama in perpetuity? What will it take to find a position that is not in reaction to an *a priori* enemy, whether imagined or real? From a symptomatic perspective, it is clear that we haven't forgiven and our reconciliation is tenuous, so we can expect that it will break, in small ways, and in big ways, to remind us of a past we try to outrun by controlling the future. We will overcome our conflict only when there is diverse participation in bottom-up public processes that cuts across classes, races, genders and ethnicities, and allows for unique experiences of sharing and new conceptions of forgiveness to emerge from the social, so that the discourse is challenged and informed at every level in society and is diversified enough to evolve. Without this, we will remain stuck in prescribed modes of communicating and acting upon issues of forgiveness, and run the danger of relegating themes of resentment and exclusion to the margins of the discourse, where they are fed by undercurrents and remain invisible until the conditions emerge for more explosive and destructive expression of it. The impossible consti-tutes a small part of the past, a greater part of the present, and the larger proportion of the future. So our main task is to think about how we liberate spaces for contemplation and expression so that we can reach for the impossible in the future, and even if we never arrive at absolute forgiveness as a society, that we can at least successfully negotiate our past when it emerges, and move beyond it on the whole.

Chapter 3

The Wakeful Dead

*The masterwork is unknown, only the work is known, knowable.
The master is the head, the capital, the reserve, the stock and the
source, the beginning, the bounty. It lies in the intermediary inter-
stices between manifestations of work. No one can produce a work
without labouring in this sheer sheeting cascade from which now
and then arises a form. One must swim in language and sink, as
though lost, in its noise, if a proof or a poem that is dense is to be
born. The work is made of forms, the masterwork is a formless fount
of forms, the work is made of time, the masterwork is the source of
times, the work is a confident chord, the masterwork trembles with
noise. He who does not hear this noise has never composed any
sonatas. The masterwork never stops rustling and calling.
Everything can be found in this matrix, nothing is in this matrix;
one could call it smooth, one could call it chaotic, a laminar
waterfall or clouds storm-crossed, a crowd. What are called
phenomena alone are known and knowable, avatars of a secret
remote proteus emerge from the clamourous sea. Visible and
beautiful are the dispersed tableaux; beneath the green serge veil,
lies the well. Empty, full, will we ever know? When there is an
infinity of dispersed information in the well, it is really the same
well as if it were devoid of information.*
—Michel Serres (Genesis, 1982)

So what of us; caught in the multiple? In the realm of the senses
we are relegated to a continuum of sampling and digesting with
the body instrument. The multiple contains both artifice and real
but provides only glimpses of both. The multiple is the domain
of the zombie, the undecidable. We, caught within this multiplicity, are also zombies; navigating the living and the dead,

within and without, through everyday existence. Within us, and without us, there are two conditions. We speak, always in part with two voices: our own voice, and a borrowed voice. We navigate the realm of experience through the superposition of two modes or conditions. As the zombie, we inhabit both the city of the living, and the city of the dead—simultaneously—albeit in varying proportions and superpositions of both.

The belle noiseuse is not a picture, it is the noise of beauty, the naked multiple, the numerous sea, from which a beautiful Aphrodite is born, or isn't born, accordingly.
—Michell Serres (Genesis, 1982)

It is a *duality*, not a binarity, of the polis and the necropolis respectively. That which is alive and present resides in the realm of the polis, while that which is dead yet present resides in the realm of the necropolis. We see the world through two eyes that our brain fails to integrate into a coherent bifocal reality; deceiving us into believing that we are awake when we are dreaming and we are dreaming when we are awake. We are only living when we are mostly dwelling in the domain of the polis— that is, dynamic, interactive, heterogeneous, unpredictable, improvisational—while our death (or even hell) is when we are consigned to the banal, mundane, homogeneous, procedural, predictable bureaucracy of the necropolis. We, trapped in the multiple, are in the process of both living and dying.

Surely, when this is considered, the notion of rigour can only be erroneously employed against the reality of the multiple, in that it acts as a mechanism for falsely reducing the irreducible.

Displacements for the purpose of seeing borrow pathways, crossings, intersections, in order that scrutiny may focus on the detail or pass into a global synthesis: changes of scale, sense and direction. The sensible, in general, holds together all senses, all directions, like a

knot or general intersection ... Visitation explores and details all the sense of the sensible, implicated or compacted in its knot. How could one see the compacted capacity of the senses if one separates them? We have visited this capacity without dissociating the senses of the word visit.
—Michel Serres (Les Cinq Sens, 1998)

Spirit sees, language sees, the body visits. It always exceeds its site, by displacement. The subject sees, the body visits, surpasses its own position, goes out from its role or word ... The body goes out from the body in all senses (dans tous le sens), the sensible knots up this knot, the sensible in which the body never persists in the same plane or content but plunges and lives in perpetual exchange, turbulence, whirlwind, circumstance. The body exceeds the body, the I surpasses the I, identity delivers itself from belonging at every instant. I sense therefore I pass, chameleon, in a variegated multiplicity, become half-caste, quadroon, mulatto, octoroon, hybrid.
—Michel Serres (Les Cinq Sens, 1998)

The multiple cannot be reduced in the absolute; that is, contrived as mono-dimensional. In the realm of the senses we experience only fragmentary information on (or of) the whole; part dream, part real. It may be consistent but it is never complete. Like Serres, I am "attempting to rethink time as pure multiplicity" (Serres, 1982). Yes, the multiple consists of noise, but within the noise are heartbeats, rhythms, movements—sustained or periodic—a thousand drums in the noise, "behold how the noise divides itself" (Serres, 1998). This is not obscurity. Obscurity is reducing the multiple to the mono-dimensional.

In South Africa the notion of nationhood is still emerging. Caught in the multiple it takes on different forms as it attempts to wrest free from waves that ripple through the multiple. Different waves yield different forms. The nation emerges in the interactions between us. Some faces slip below the wave to make way for others, framed by the question or hypothesis.

So how do we engage the multiple with this knowledge? Every question, every hypothesis, every interpretation cannot escape that it straddles both worlds; the polis and the necropolis. We are always caught in this indivisibility, and so we must learn to see with both eyes and to detect both tongues. Hot and cold is our constant condition but the result is not lukewarm. It is not so much a mixture as a superposition; a simultaneity. Even as I write, the pen is writing too. There is no separation between subject and object in the multiple; especially the multiple as engaged through acts of creation and co-creation. The necropolis is lukewarm, numb, devoid of feeling; where the constancy of work and labour, production and consumption are flows cast adrift in a sea of nothingness, living to accomplish only survival, and not existence.

Separating 'what was' from 'what is' and 'what will be' isn't easy. What marks them as they emerge within us? What kind of awareness does this call for?

We have to go beyond the rational to negotiate the impossible and create or co-create the future. We have to find and negotiate the elusive flows, be sensitive to rhythm. Behind all interactions lie strategies. We adopt different strategies in different interactions (Cilliers, 1998). Like heartbeats trapped underwater we beat out different rhythms as we interact, and it determines how we dance out our interpretation of reality in relation to each other. Strategies, like heartbeats, collect until they become movements of a dance. Either we can feel each other, or we can't and remain out of sync while the flow passes around us, but not within us, not through us, and we remain outside of it, sometimes by choice. A strategy behind every interaction, or several, switched and interchanged through the interaction. A strategy isn't fixed to itself; therefore it does not determine the outcome, but provides clues as to how things may—or may not—unfold. Strategies inform every interaction but they do not dictate the collective outcome. They may rival or coalesce as the context dictates. That is up to how their rhythm meets that contained within the noise; whether creation or destruction is at work in the symphony.

But while strategy is about rhythm, it is also about position. The most powerful position is that adopted at counterpoint. Counterpoint exposes the gaps where new rhythms can be established, or momentarily explored. We need a sense of rhythm in order to find counterpoint. The movements of a boxer, or a jazz improvisationalist, are based on the same understanding. History cannot be ignored, but the moment must be seized in order for the improvisationalist to vanquish over the prevailing rhythms and project a clear voice amidst them. The improvisationalist must know the rules in order to break the rules, but the manner of their actions illustrates their personal mastery or clumsiness in the process of creation, innovation and revolution. The strategist must dance to different rhythms, like a philosopher tasked with negotiating the impossible alongside the possible and the probable, so that past, present and future are simultaneously negotiated with the appropriate awareness.

Philosophers are there to muddle what needs muddling and simplify what needs simplifying. Identifying which is appropriate is the essence of mastery in this regard. Mastery cannot be taught. It must be discovered, and then verified through its ability to strengthen or subvert and change. Mastery hears its own heartbeat and the heartbeat of others simultaneously. It is not an integration but a superposition. Each coexists, like different frequencies of a spectrum; distinct but continuous. In a temporal moment all frequencies are present, but what is seen is determined by the filters that are used. Mastery filters and unfilters. It attaches and detaches appropriately and finds counterpoint in between; choosing to exploit it only when the appropriate time arrives; complete with its context, emotion and sense.

Chapter 4

Resurrecting Lazarus: Leaders, Philosophers & Strategists

Lazarus, untethered, without mastery, falters and dies, but can be resurrected. For Lazarus, death is but one of many positions amongst the multiple, so he emerges only when the master calls. And that is what philosophers, strategists and leaders do; they pronounce the call to a higher order of things. Leaders who appeal to the lower order tend to perpetuate sameness and mediocrity under the guise of difference and heterogeneity, and trap us within machinery they understand, but we do not. But when Lazarus awakens the call is undoubtedly that of a master that brings him alive. He cannot awake from his slumber unless revolution or innovation arrives, delivered by another or by his own meanderings through slumber, but fundamental in its ability to unlock the enduring daze of ritual and repetition. He emerges as a phoenix only when his heartbeat rises above the water, breathes oxygen unencumbered and beats loud, for everyone to hear. Otherwise, he is lost in the noise, indivisible from the flows around him; unable to raise a signal in the Brownian milieux he persists, head down and brow furrowed along paths already tread, destined to find no release.

Lazarus the zombie becomes a phoenix in the presence of a master because the call gives him life, awakens him from slumber and creates new spaces for growth to emerge so he can live anew and expand from his position to occupy higher territories. That is, when the master is of his choosing. When the master is imposed Lazarus remains lost within the necropolis, unable to find true expression, true connection with the world of the living, unable to rise above the "green serge veil" (Serres, 1982). Lazarus seeks to become his own master, and to do this, he seeks out others to help him find his own voice. The voice that is distinct within the noise—even for a moment—so he can become the eternal improvisationalist, at one with the rhythms of his own existence and that of the multiple. Otherwise, he seeks only survival, whether imperial or pauper, in provision of needs and rituals of pleasure. Life without rhythm is like food without spice; essential for its character even though no nutritional value is gained from it. That is, it is empty of functional value (except perhaps the ability to preserve) but there is no product without it.

Lazarus needs to empty his cup to the moment, silence the voices within and find his own voice. Amidst the confusion of internal and external voices, Lazarus seeks out guidance. The master is not a rescuer. The master knows that intervention and disruption cut across time, while observation is trapped in the past and the present. We cannot observe the future. Intervention and disruption, though, address past, present and future. The lords of scientific pragmatism and reason remain arrogantly sane amongst the madness of the multiple, but are often shown to be fools. The masters are different. They wade through the muck unperturbed, but reflective. They eat up uncertainty. They digest context and regurgitate and excrete context. The leader as master—in the sense of having achieved mastery, that is, as warrior, philosopher, facilitator, politician, humanitarian and environmentalist—is the key to a successful future vision of society.

The master is keenly aware of shifts in context and nets and discards with care, adapting his strategy as each interaction unfolds. The master must maintain a heterarchical perspective and a rhythm that allows him to engage with externalities on his own terms. When does Lazarus become his own master? Surely when he has undertaken the call wholly and fully and so reaches his own understanding through the process of negotiating change. The master and the call are just facilitators, to be abandoned like a raft which has no purpose once a river is crossed. Lazarus has to do the work.

The master facilitates through disruption and intervention of scripts or cycles that are held in place by false notions and perceptions, and play out repetitively, keeping Lazarus trapped in the necropolis. He is able to find the counterpoint in Lazarus's rhythms, and so intervenes and disrupts to destabilise his beliefs. The master may be artful or brutal in his approach; that is, as the context and nature of the interaction dictates. The master, in doing so, helps Lazarus identify rigid patterns that are constructed from false beliefs; thereby opening up spaces for growth and helping to establish new rhythms and flows from which new trajectories and behaviours unfold.

Coming unstuck from mindless repetition of patterns of thought, emotion and sensation is the purpose of disruption and intervention. That is, to shift Lazarus from the rail or track he has become confined to. Like Sisyphus raising the boulder to the top of the hill—trapped, maddened, yet unable to see or accept the changing context—Lazarus remains stuck in the neo-liberal consumer dream, always with a sense of belatedness; of always arriving late at predetermined destinations and on well-worn tracks. If he stops engaging with the cycle then hope emerges from within the imprisonment of the necropolis.

Chapter 5

Lazarus and the Traveller

And what I've realised about life, is that we are simply here to witness it.
—Mr Mok

Travelling is the urge to witness life so that its stories can go on forever. We collectively generate social reality, whether we are here, or long-gone, dead and buried, ashes. Even those who are not yet here, by existing in our imaginations, inform the reality — *the multiple* — in which we exist. It is assemblage; not layered, nor flat, but evolving heterarchy, held together by our collective engagement in and with it. So, is everything remembered? Well, perhaps not explicitly, but definitely implicitly to some, often varying, extent. The 'memory' of social reality — a 'field' co-mingling and co-evolving with the assemblage — is a duality; a collective psyche that constitutes a continuum between the conscious and subconscious. Even things, long forgotten in historical memory, may still be with us today and may manifest in varied, subtle forms.

Viewed this way, any perfect understanding of society is impossible. It is futile to construct a model of it to be able to control it. Better to have a model that helps you navigate it, imagine it, and act upon that imagination to contribute to creating — or co-creating — it. Trying to understand it, purely for the purposes of prediction and control, fails to acknowledge the real nature of social reality, and does not equip us to negotiate it and transform it. Whatever we see and experience enters this reality through us, through how we exert our psychologies within it, and through the stories we tell.

Even our sciences, mathematics and statistics are interpreted through stories; narratives of how we 'model' what we observe. They just deploy stricter languages to construct these stories or 'theories' of reality. These 'stricter', more constrained languages may help us with prediction, but this is precisely because they attempt to cast away the very elements of spoken language that enable it to negotiate complexity. Hence they can predict what may happen with simple and complicated systems, but they cannot cope with complexity. They do not have the language—or elements of language—to support an understanding of, and artic-ulation of, the complexity of social reality.

And so we must travel through all the realities we can muster, so that we can internalise the multiplicity of it all within us, and reproduce it from us back out into the world, thereby changing it by being within it; the 'material' and the 'virtual' both make up the real. Travelling is the pilgrimage through the vast oneness— "the multiple"—as Serres put it; "variegated, half-caste, octoroon", "beneath the green serge veil", but it is also an action that broadens reality through the reflexive 'processing' of reality and remaking or reproducing of it. A single agent, network, cluster or organisation of agents may exert back small, seemingly insignificant, or large influences on this reality.

Whether there are many universes which we simultaneously occupy or not, it is beyond dispute that many realities constitute the social, and society. We generate it, observe it, live within it and drive changes within it through our sheer heterogeneity. *That* is what makes social reality complex. The more we witness, the more we both reinforce and destabilise it at the same time. This is what brings social reality to life; what energises it, gives it character, and allows us to imagine our engagement with it. It is what makes us engage with it as a 'whole' of sorts, as something more than the sum of its parts. Ironically irreducible, but recognised as a whole unto itself by all those within it, it is personified into wholeness by being granted the attribute of identity.

So what does all this mean for how social reality changes and transforms? What can we make use of—in respect of how social complexity manifests within social reality—to influence society to bring about the futures we desire for ourselves, and those who are not yet here?

To the traveller, there is a deeper guidance system; an awareness of how witnessing transforms them and the world along with them. It is a more passive medium through which change is effected. What can we learn from the traveller's strong engagement with the multiple nature of social reality?

The traveller is acutely alive, very far away from the zombie-like existence of Lazarus, lost in the multiple. The traveller is awakened to new realities every day, and journeys through them purely because witnessing them keeps the traveller truly alive. And with that travellers become reproducers who broaden the general social reality; pushing its boundaries through extending the memory and experience that is garnered through travel.

The traveller awakens us to the inevitability of the hetero-geneity of social reality; that it can only ever increase as we increase and become more connected. Only catastrophe can divide us again, and return us to the fragmented existences that we 'supposedly' existed in when we were few. And so there is something to say about Professor Judt's observation that social homogeneity historically correlates with high levels of mutual social trust and a stronger social contract (Judt, 2011). How well does it predict the plight of the future, our future?

The social stability of homogeneous societies becomes metastable in an increasingly heterogeneous world. In more closed societies that are internally self-reinforcing their values, beliefs, norms and behaviours, small differences are amplified, whereas in more open, heterogeneous societies small differences are by necessity tolerated, encouraging diversity and a broader range of options for society's evolution and transformation. Divergence and convergence become equals, with more of both present, and in contestation with each other, in heterogeneous societies.

Both the material (i.e. physical, spatial) and virtual (i.e. non-physical, cyber-dimensional) spheres hold the key to broadening the sphere of the individual into multiplicity. They are both assets and obstacles at the same time. An inland herder or coastal trader in Somalia can now access people, social networks and transactions that lie far outside his or her locale and scope of daily interactions, through which daily survival and social life can be enabled. This multiplicity is increasingly a fact of life, because even the stationary individual can be travelling through vastly broadened spheres of interaction, where far and near, past and future, all become present, immediate.

In the long term, this increasing multiplicity of social reality has the potential to become a generator of a new stability, one that is negotiated through increasing tolerance of difference— and perhaps 'différance'. A new social reality where the traveller becomes the norm, and the stationary become the exception; a reality where everyone travels through different spheres and internalises multiplicity through engaging with it for themselves, directly; albeit more or less filtered and unfiltered. This kind of reality, currently understood as the exception—i.e. in historical terms—is increasingly the most likely prospect for a new social reality. It is one that we must both make, and internalise, in order for it to work for us as societies.

We can no longer avoid engaging with the changes that are unfolding in the world we live in, and must prepare ourselves for a new social bond of trust, a new social contract; one that is forged on difference, and not sameness. One that absorbs multiplicity and reproduces it to its benefit, not to its detriment.

Chapter 6

Obstacles to a Third Way: Ideology before Analysis!

When ideology precedes analysis, it in effect serves to negate analysis. Analysis then becomes constrained by the boa constrictor of absolutism and ideological 'canonism', and a hierarchy is imposed upon analysis that locks it into a tauto-logical double bind or catch-22. To put it simply, if you put the cart before the horse you shouldn't be surprised that you end up in the same place all the time.

The 'horse' of analysis becomes restricted to a few steps forward and a few steps backward. The 'cart' of ideology or absolute theoretical hegemony becomes the pivotal axis and tether that governs the direction and extent to which analysis can go. The horse, confused, pulls backward and side to side, finding relief only when it remains in one place. Every direction is fraught with tension, so the only option is to remain direc-tionless. In this way, analysis always returns to the same place, because ideology has conditioned it to be bereft of agency, its inquisitiveness and imagination put asunder for a 'greater' purpose.

Theoretical frameworks, if they are to remain honest—i.e. not necessarily truthful in the sense that they make the claim to absolute truth but honest in their approach to the subject matter under analysis—should be deduced, or even abducted (see Chapter 12, "The Multiple as Filter"), but never inducted without great care. That is, induction from theoretical frame-works in complex, real-world contexts usually constitutes a complex fabrication. It is fabrication because it pretends to derive that which it is actually premised on.

Real-world contexts are complex, and their behaviours cannot be inducted from a model, theoretical framework or pure methodology. Scientific induction is only credible in systems that are simple, or at most, complicated. Inducted principles can be derived from systems that are generally simple enough to be tested by hypotheses and repeated. Induction cannot negotiate the complex. Only if the system is not dynamically changing its fundamental conditions and constraints on a timescale that negates induction, is induction then possible.

In other words, induction only works for simple, well-constrained systems that progress in a linear fashion where change is incremental and predictable. As soon as the system begins to progress in non-linear leaps and bounds, and in different directions, induction becomes less tenuous as a methodological foundation. Social, political and economic systems, being fundamentally social in their conception, are complex, reflexive systems. They are in the domain of the deductive and abductive. Understanding has to follow the unfolding behaviour of the system. By definition, it cannot precede it, except through what can only be termed 'oracular' insight.

And ideology precedes analysis regularly in a global society that constantly looks backwards to its historical foundations to answer questions about how to face the future. Ideology, to be sure, is by and large the starting point of many debates, analyses and opinions that are generated over the question of which socio-political and economic systems are responsible for the global crisis we have entered into. And it is not simply a crisis of economic systems. It is a crisis of how to move forward, and to generate new ideas about how the global polity and socio-economy can be best positioned in relation to each other. As articulated by Žižek (2011), "the field is open" and the marriage between capitalism and democracy has ended, yet we are struggling to find a new way forward.

Our imaginations and our ability to inquisit have indeed become conditioned by ideology. Instead of maintaining a critical perspective on our prevailing ideologies, we lapse into analysis that *a priori* draws upon the ideological foundations that we prefer. We therefore become 'stuck' in self-reinforcing, circular patterns of analysis that take us nowhere. More often than not, it simply becomes analysis that searches for where to place the blame; as if any of the systems that constitute the current global order can be regarded as entirely blameless in the first place.

Yet there is more to this kind of analysis. It is analysis that is explicitly oriented around an ideological foundation, and is implicitly constructed in reaction to the opposing ideological foundation (i.e. its 'metaphysical' opposite). The analysis therefore ends up being a de-facto debate between polar opposite positions. Indeed, this is the very critique that deconstruction offers of the general methodology by which philosophy itself is constructed. This raises the question of how to embrace analysis that does not fall at either side, but that begins 'in the middle' so to speak. This is not a trivial question, for arguments that proceed from the poles are the fundamental obstacle to generating what may be regarded as a 'third way'.

In other words, what kind of third way can be 'constructed' or 'formulated' (these terms are unfortunate, so they are used here with reservation) if we do not start from the edges or the poles, but start from the middle instead? Note, I do not mean to start from an ideological middle ground, but from a middle that makes observations of the poles, and remains acutely aware of them and how they influence us. There are two aspects that emerge from observations that are made in the middle that require closer inspection. Firstly, that there are fundamentally irresolvable 'undecidables', i.e. the decisions, phenomena or events that fall between metaphysical opposites are fundamentally irresolvable, and cannot be avoided in political decision-making (Derrida, 1992). Secondly, that the polar opposites, when they approach the extremes of either end, begin to mirror each other, resulting in a different kind of obfuscation, where it becomes difficult to distinguish one polar opposite from another (Žižek, 2009).

The *former* observation is made by Derrida, in his careful identification of the 'undecidable' in political decision-making, where he concludes that any political decision that forgoes the "ordeal of the undecidable" is not in reality a political decision but, rather, becomes the mere "unfolding of a calculable process" (Derrida, 1992). That is, there will be fundamentally irresolvable, 'undecidable' factors that occupy the territory of the middle, otherwise referred to as the 'logic of the included middle' by Plato (Max-Neef, 2005). These irresolvables define where metaphysical opposites differ from each other, and resolving these undecidables requires more than theory. It requires learning, participation, integration and negotiation, i.e. strategies that engage directly with the contextual specificities that bring about undecidability, and which can possibly lead to the resolution of these undecidables under certain context-specific conditions. In this way, undecidables may end up being resolved in context (though not always), yet remain unresolvable at the broader theoretical level that is removed from the specificities of context.

The *latter* observation is made by Žižek, when he identifies how the right-wing conservative tea-party movement resurrects the rhetoric of the labour movements that existed 50 years ago in their conception of the tea-party identity as fighting for the rights of the ordinary worker against the irresponsibility of big government and big capital. The same is true of 'left-wing' eco-movements that upon closer examination are mainly biocentric in their values, beliefs and norms. This biocentrism itself hails from the historical support for conservation biology efforts that saw predominantly indigenous peoples being herded off their own lands and into reservations, so that the 'pristine' natural environment could be maintained, devoid of human influence (hence biocentric). That is, eco-movements whose foundations are in fact profoundly anti-social have staked a specious claim over the socialist territory of the left. Only in Latin America and India have eco-movements become profoundly social in their approach, warranting a clear membership of the left.

Perhaps another dimension of analysis can be added to this: primarily that as those at the extremes of the poles increasingly embrace their ideologies with what can only be described as a messianic zeal, they begin to resemble a church or a cult (i.e. an institutionalised set of beliefs that are taken on faith), where the only distinguishing factor between a church and a cult is the size of the following. Where ideology becomes the unshakeable foundation from which analyses are made — instead of being regarded as a hypothetical framework — it follows that the values, beliefs and norms that govern the analyses or debates are also unshakeable (if deftly hidden) preconditions that impose a hierarchy upon the analytical framework, which ceases to be interpretive or reflexive at this point. This unquestioning 'lock-in' to foundational values, beliefs and norms is not irrelevant. It is the main obstacle to going beyond a bipolar theoretical debate because dogma is followed by rhetoric at best and sophism or sophistry at worst. Moreover, behaviours are founded upon values, beliefs and norms, so real-world changes in behaviour are obstructed at the same time, i.e. the foundation for action is also subverted, or at the very least obscured.

The consequence of ideological fundamentalism is that we are unable to effectively bring about the necessary changes in the way we think, that is, to produce the substrate that may ultimately result in changes in actions and behaviours that govern the global political, socio-economic and ecological 'condition'. And this 'condition', as a result, remains unchanged. To put it forcefully, the pro-free market and anti-free market ideologues have created their own respective sacred grounds that are too hallowed to question, and become external to analysis by being implicit in their interrogative frameworks.

In a very genuine sense, debates of this ilk resemble a debate between parrots; more often than not they can only talk past one another as their 'vocabularies' are largely pre-fixed, and no real meaning is engendered in the interchange. They celebrate the undecidables between them as signifiers of their ideological purity. Yet at the same time these debates resemble a debate between sophists or conmen, as they appropriate the arguments and critiques of each other where it 'fits in' with their ideological frameworks, and speciously lay claim to them as their own. The latter indicates that a postmodern relativism is erroneously employed to subvert the claim that each hold to their own precious beliefs. That is, not only do they differ violently, they also attempt to appropriate the ideological territory of their opponents at the same time. It is truly a war, in which any and all tactics are considered fair.

And predictably, they despise any and all who occupy the middle. Indeed, in this they are not far away from the Christian God, who proclaims that only those who are passionate for God can be accommodated within the halls of the Church. The 'lukewarm' will be spat out (Revelations 3.16). Those in the middle are not just spat out, they are spat on, literally and verbally. Yet it is not on the basis of analyses that they are spat upon. It is on the basis of concretised values, beliefs and norms that are taken for granted as foundational. Nothing could be further from the truth, however, as the failure of both positions demonstrated towards the end of the previous century. It is true that absolute objectivity is also an impossibility, but where the values, beliefs and norms that are subjectively held become implicit and unacknowledged—i.e. no awareness of them exists except in their virtue as absolute and non-negotiable precondi- tions—they become constrictive to the generation of any honest debate or analysis. They are therefore oblivious to their own strategic orientation, because it takes the foundations from which it proceeds as self-evident.

That is, the dishonesty is a kind of ignorance that prevails to the ultimate end, discrediting all and sundry around it that doesn't make the effort to fit itself into a particular framework or position. It is a narrow-mindedness that is holding us back in our quest for a new way that will rescue us from the conditions of polycrisis and global hegemony. Ideas and new frameworks that fall outside of the mainstream poles are largely ignored or despised as compromises and cast aside. Perhaps it is the hegemony of analyses and ideas that is proving the most difficult to break, while the hegemonies of power and wealth have found themselves floundering in the winds of change that the beginning of the 21st century has brought with it.

Note: This chapter has been adapted from a blog post that was published online on 2 November 2011; see Peter (2011a).

Chapter 7

Left Meets Right

This chapter is a reaction to two dominant perspectives that, ironically, mirror each other very closely. Firstly in reaction to the emergence of a right-wing discourse that rails against big government, and places all its faith in 'the small guy' who, in a neo-liberal fantasy world where all things remain equal, will inevitably triumph over restrictive regulation and taxation and prevail in the 'free market'. Secondly, it is a reaction to the emergence of a left-wing discourse that *a priori* assumes that the mechanisms of global hegemony are at work in every event and circumstance that unfolds upon our television and computer screens, fed by a corporatised media that slavishly and un-objectively executes the 'will' of their corporate owners. I will not debate the merits or the facts of the respective positions adopted by the left or the right. Instead, I will point out their ultimate similarities, and reveal how they become mirrors of each other, ultimately hijacking the political space for free expression and political debate, condemning those who do not fall within either camp to the obscurity of political apathy.

The first view, maintained by the right, relegates any taxation, regulation, etc. as evil instruments of the left, who are constantly trying to find a way to dupe or con them into accepting dressed-up forms of socialism and communism into their hallowed freedom, and to distort 'true' (read neo-liberal or neoconservative) democracy. Fox News—the Rupert Murdoch-owned channel that is unashamedly pro-conservative—is perhaps the best example of this view, where all debates are located in frameworks where the 'right' (in the sense of 'correct' here) view is presented as self-evident and obvious. It is used as the framework for judging dissenting or differing opinion. It is not a neutral ground where opinions are debated on their own merit. Opinions are filtered through the 'tea party' discourse, chewed on like cud, and either expelled or grudgingly swallowed in the name of upholding the tenets of 'freedom of opinion'. The conspiracy mill is very much alive and constantly at work on the right, rooting out the schemes of the left, finding fault with their nefarious 'do-gooder', bleeding-heart nonsense that is used to further an agenda of inciting class warfare and, to put it simply, to use higher taxation and state resources to keep the lazy and unproductive afloat. The right, of this ilk, are not open to critique. They already 'know' what the problem is, and what the solutions are, and have no need for debate except in demonstrating that the freedom to debate still exists. This is ironic, even paradoxical, as it means that the freedom to engage or act does not exist in reality. What exists is a shadow of freedom that masquerades as the real thing. You can disagree, but in reality, nobody is listening for disagreement. Their minds are made up and they only seek out agreement.

Likewise, the left have slowly but surely perverted the discourse on global hegemony into an *a priori* framework that informs all their analyses and opinions on global or local events. No longer does the left make recourse to objective analyses before arriving at an opinion or idea of what is transpiring. Rather, the discourse on global hegemony has gained so much strength that it serves as a conspiratorial meta-narrative through which all events and situations are filtered in analysis. There is a clear evil, and a clear good, and never the twain shall meet. The West, driven by its corporations and governments, is intent on dominating the resource bases, polity and social values of all those within its borders and those outside of it. It is an evil machine that churns out self-serving platitudes to purposively dominate all events and all situations across the globe. The Western Empire trundles forth behind its massive weaponry, bombing whoever stands in its way, making a mockery of the very values it purports to uphold, not unlike the hitherto Catholic Church of yesteryear. Any action it takes is not to be trusted, as it is loaded with hidden agendas to control the rest of the planet and its resource bases. In return, the left cheers any and all who make the appearance of 'standing up to' or 'bucking' the West, and regards them as brethren to which their loyalty must surely be given under any circumstances. They are loathe to criticise 'one of themselves' because they are all brothers and need each other in the great 'struggle' against the West. They are reluctant to be forcibly critical of the Chinese and Cuban governments, or Gaddafi and other 'left-aligned' populist dictators, despite the availability of evidence of human rights violations in these countries. Like their right-wing counterparts, the left also believe that there is no need for debate, except as a show of their appreciation of the need for 'freedom of speech'. Yet they are not listening when opinions go against their framework of beliefs, as their beliefs have become foundational. They are not up for debate. Rather, their beliefs become a standard by which the debate (and those of dissenting opinion) are judged. Often this judgement takes the form of "either you understand or you don't", and the left resorts to clichés such as "there are none so blind as those who will not see!" to justify where it stands. You have to be able to 'see' the conspiracy by adopting a particular framework of interpretation; the interpretive framework does not necessarily have to follow from an analysis of each situation, taken on its own merits.

Yet there is something central in this attitude, because it is shared by both the left and the right, and in this sense they become mirrors of each other. It is political mysticism. It is simple to diagnose if you do not fall completely within either side. It is that there is a shared belief that understanding can only be achieved through one set of filters, and that these filters can only be understood or 'seen' if the individual searches deeply within themselves and interprets the evidence 'honestly' or 'in the right way'. It is mysticism dressed up as objectivity; a claim that there is 'correct' way of seeing, to the exclusion of others. Both sides appeal to a mystical sense of 'knowing' or being 'in the know', in order to arrive at the positions that they prefer to arrive at. In both cases, circular logic is imposed, i.e. they begin from a foundational view, interpret all evidence within that foundational view, and thereafter declare the foundational view correct or justified. It is tautological; it starts and ends in the same place. Causality isn't drawn out from the specificities of context— rather, causality is framed within the original, foundational position and over-arching meta-narrative. It is not true analysis. It is the same mystical belief shared by cults, who state the same thing; that is, "if you can't see the mystical truth, then something is deficient in you"—it is your fault for not being able to see things their way. It is not incumbent upon them to convince you, you have to earn the privilege of sharing the right to the hallowed 'truth'.

Both the left and right have, in their extreme polarisation, become cult-like conspiracy mills in which any objective analysis of an event that occurs in real time is obscured by historical meta-narratives that claim a hegemony of their own over all present and future events. Yet seldom do the proponents of either side step back to ask the critical question, "if we already know the answer, then what is the point of analysis?" And this is the question that reveals what the right-left mirror constituencies have become—they have become fundamentalists who believe in foundational views. Things as they are, or will be, have no uniqueness, they have no 'contextuality' to those who occupy the poles. Rather, they can only be understood through foundational filters.

And this is where the problems begin, for the exclusivity of these cults relegate a large number of people to the domain of the politically apathetic. Why? Because apathy is not just a consequence of laziness or political illiteracy, it is also a consequence of the exceptionalism of the right and the left. It results from perceiving that one is unable to engage politically, except if one adopts one of the sides in this bipolar form of democracy. In short, starting in the middle is useless. You have to choose a side. For many, they become apathetic because the moment they express an opinion, either side pounces upon them and rebukes them for not knowing the mystical 'truths' that they should know about 'how the world really works'. And nobody likes *not* to be 'in the know'. So shutting up is the first option. The second option is to regard both sides with the same contempt that is reserved for religious cults and to block them out completely. They ruin life by harping on the same chords and singing the same songs, over and over again. In turn, the left and right predictably withdraw to their self-righteous thrones and denounce the state of knowledge and understanding amongst the 'masses' in the world, who are deceived by the 'powers that be' and cannot distinguish right from wrong. Very scarcely do they stop to consider the thought that this disengagement, this apathy, might actually result from the polarisation of the right and left, which leaves the space in the middle a political 'no-man's land'.

To paraphrase the comedian Chris Rock, "Everybody's so busy wanting to be down with the gang … be a fucking person … *listen* [to the issue] … let it swirl around your head. *Then*, form your opinion!" It is this gang or cult mentality that has infected the global political state of debate. It leaves a huge, yawning gap in the middle, where the possibility of finding a third way actually exists. Yet without any legitimate free debate being allowed to grow and mature out of this middle, no new ideas are being born. We oscillate in an endless debate between outdated theories that were written in an age that has long past. It is almost as if we are dragging the past into the present and forcing it upon a new age and era because we have run out of ideas. Instead, all we have is our suspicion of 'the other' to draw on, and those who dominate the political spectrum love this as it plays perfectly into the divide-and-conquer strategy that makes it easy for politicians to dupe us. It prevents us from formulating a system that can eventually displace rhetoric, and absolves us from dealing with the realities of poverty, inequality, oppression, hunger, slavery, war, etc. in any substantive measure. Who needs facts and textured contextual analysis when the moment you open your mouth, an 'angel' sings the truth through you? What more could you ask for?

Real freedom, according to Slavoj Žižek in "First as Tragedy, Then as Farce" (2009), is the freedom to fundamentally change the political systems we live with. This is a view that appeals to my sensibilities. The freedom we have today is the freedom to "live as if we were free" (Žižek, 2009; paraphrased), and Žižek diagnoses correctly that both the left and the right are complicit in landing us in this mess. In the absence of any substantial progress on agreeing what real, fundamental changes are necessary, we are relegated to the domain of the mystical in our politics. We either 'understand' or we don't. Anything else is anathema. The rhetoric of the left and the right becomes ineffable—it needs no authority but itself—it becomes the word of God. This bland, but dangerously polarising, political mysticism ultimately undoes freedom by undoing the secularity of freedom itself from within, but employs dogma and rhetoric instead of religion. The foundational views of either side are elevated beyond reproach and are not up for question. It is my contention that this is why we are failing, over and over again, even 14 years into the new century, to formulate a new politics. It is because we are not truly free. And this time, it is not governments who are robbing us of our freedom, but our own propensity for recreating ideological no-go zones that have come to define our lifestyles, personalities and, generally, our reason for existence. In short, we have become confused between the freedom to change systems and the freedom to champion a position. The freedom to champion a singular position eventually weighs down like an anchor around our necks, and we become stuck, rigid. What kind of freedom is this? A 'groundhog day' freedom is not freedom. It is the farcical enactment of freedom.

Real freedom allows and enables us to let go of cherished ideals and beliefs (even for a short while), to be able to entertain perspectives from the viewpoint of others. If we cannot achieve this then the other remains a monstrous construct, to which the only logical end is the use of force or violence (at worst), and exclusion (at least). Where there can be no dialogue, the only recourse to resolve conflict becomes violence, which is easier if the other is perceived as your exact opposite in the first place (i.e. an abomination of you). The left-right mirror opposites are like matter and anti-matter. They emerge out of a vacuum, moving backward and forward in time, and then they collide and annihilate, disappearing into the vacuum. Nothing new is created by their emergence, or their disappearance; energy is conserved in the reaction because they are equal opposites of each other. Their entanglement does not result in the creation of anything but their mutually brief existences, and then they are gone, unable to be grasped or drawn into the real. It is the very anti-thesis of the Gandhian approach, or the approach taken by luminaries such as Martin Luther King or Nelson Mandela. In this new world, 'love thine enemy' is utilised instrumentally, in much the same way the rhetoric of 'free speech and debate' is invoked and the farce of 'objective analysis' is made. No real effort is made to interrogate the fundamentals upon which either side stands, by themselves. Instead, what we observe is two cult-like followings that cast barbs at each other and stand in judgement of each other. Not everybody has the stomach for this, and the level of disengagement and apathy towards local and global politics we see in the world today is a consequence of the failure to allow for true pluralism, where freedom reigns. Rather, politics is led by the exceptionalism and exclusivity of the cult, and if you haven't had your mystical moment of realisation then you don't qualify to enter the political domain. Indeed, you are worse off than an island in no-man's land. It is for this reason that my position is that it is ultimately better to be hated by both the left and the right than to be loved by them, because if you allow them to love you they quickly rob you of your voice. You end up speaking with someone else's voice, and not your own—that is; the voice of whichever visionary, angel, God or demi-God the cult chooses as its soothsayer at that moment in time.

In the spirit of this piece, you, the reader, have all the freedom in the world to disagree with me. This is not a search for yet another absolute position, but an appeal for greater tolerance, broader debate and for sincere analysis that can help us formulate different ways of enabling our freedom in this new global era. My appeal is that you seek more deeply to understand those positions you might disagree with most. It is not easy to do this, but it does help open up dialogues and debates that have a genuine freedom and not a contrived freedom in which agendas and positions are already worked out. My appeal is that you judge the transition that the globe is undergoing as vigilantly as you can, but allow the context and evidence to inform your judgement, rather than retro-fitting the evidence to a framework you have already decided upon. The more you fail to achieve this on the left and right, the more you alienate those who aren't 'believers' and rob them of their chance to enter the debate. It is your messianic exceptionalism that catalyses apathy. Ultimately, however, I am acutely aware that the majority of you will end up hating me for diagnosing you as part of the problem. I, being hated by both sides, also become part of the problem in this respect, but I welcome your hate. It is nothing new to me. I'd rather you hate me than you love me, because your love is more dangerous than hate.

Note: This chapter has been adapted from a blog post that was published online on 22 October 2011; see Peter (2011b).

Chapter 8

Necropalyptic Politics:
The Politics of the Dead

Necropalyptic politics results when individuating, bureaucratic, alienating and impersonal forces infect the polity, economy and ecology of society, bringing about fragmentation in their relationships to one another. In this fragmentation, institutions render themselves as reluctant, self-serving paper-pushers rather than effective actors that contribute to advancement, improvement and change in society. We need to understand change, from the global to the local, in order to adapt, intervene and disrupt appropriately.

In demanding our assimilation into neo-liberalism, consumerism and pirate free-marketeering their presumption is that there is a hierarchy of development to which we must subscribe, or remained doomed forever. However, it is precisely this assimilation that dooms us, as we flail about helplessly without our own historical, new and emerging survival tools. Requesting or forcing assimilation ironically increases our marginalisation.

True pluralism is what I am arguing for, not nationalism, not afrocentrism, because these are all artefacts of colonialism, Eurocentrism and Americanism. Pluralism allows for depth and breadth in context to be uncovered, and by teasing out paradoxical, irreducible and undecidable understandings and views we liberate the spaces of the *new*, enabling them to emerge. We can negotiate the probable, the possible and the impossible. We can find our own way.

The diabolical nature of British, American and European relations with the rest of the world emerges from their abuse of instruments of power to literally 'force' us into prescribed trajectories that have rendered our societies to Sisyphean cycles of creation and destruction, of hard toiling for an apparent stability only to lapse quickly into instability. The real message from the backers of the World Bank and IMF is: don't ever dare to challenge the framework of development as prescribed. If you do dare, then Robert Mugabe will be your only friend!

But what if we dared to assert our own agency, even sovereignty? What could we do, without violence, to send a clear message to the ageing and weak grandfathers of humankind and the world? One radical step would be to refuse development aid as an unspoken barter for lifting the trade and exploitative structural adjustment regimes that have in part been responsible for the fracturing of our societies and the ruin of our ecologies.

There is a multiple debt that we incur for allowing our so-called 'partners' to dictate the discourse on our development. We pay accordingly through economic splintering, social fragmentation and ecological loss and degradation. We pay over the elements that should, and can, make us resilient, and on our own terms.

Trapped forever to chains of aid-debt and resource-stripping we become 'locked-in' to singular trajectories of second-class citizenship in our own countries. We give priority to the same 'international' interests that have singlehandedly contributed the most to the extreme losses that most of the ex-formally colonised world has experienced. In the Sudan and much of the Sahel people kill each other over a scarcity of grazing land and cattle that has resulted directly from desertification induced by emissions from North America that land on the sea surface in the Gulf-stream, raising its temperatures above the critical limits required for human sustainability in the Sahel.

We get on with the business of killing each other because even if the causality is clear, the powerful decide whether they are guilty or not. They need no rationale except their own. Like a spoilt, overindulged child who is made to believe that the world

revolves around their needs, every little sacrifice is a victimi-
sation that they are allowed to cry over, fight over and contest
through tantrums. Beware the child of tantrums, as it is this child
who operates only by manipulation that is justified as—and can
only be justified as—self-interest. This self-interest, ironically, is
the same that emerges amongst the desperately marginalised
under conditions of social, economic and ecological breakdown.

Maybe the so-called 'first world' consists of a desperately
severe marginalisation within its own societies, economies and
ecologies that it hasn't discovered yet, and will only discover
when their mundanely secure lives and livelihoods are
eventually threatened. What becomes of Europe, Scandinavia
and North America when they are truly tested? War and
genocide are never far removed from the superficial veneer of
civilisation that the West celebrates. World War One, World War
Two, Kosovo, Bosnia-Hercegovina: these are hardly ever
acknowledged as symptomatic of a deeper, underlying sickness.
Instead, they are regarded as temporary aberrations in an
otherwise perfect system. Yet this elevated sense of superior
Western values is difficult to countenance. It is not difficult to
imagine that if European countries were subjected to the same
trials that African countries undergo, they would turn on each
other and slaughter again without compunction. The EU would
quickly fragment into short-termism and narrow-minded provin-
cialism. It's hard to believe that Europe's so-called multicultur-
alism would survive the same tests as its ex-slave colonies have
faced. They would fall down precarious social trajectories if their
wealth and influence were diminished. At its core, excessive self-
individuation is a dangerous obsession. It is hubris; it is not a
truly tested confidence and it unravels in the same way that all
messianic hubris does—it explodes or implodes. It never goes out
gracefully. It destroys the fabric around it in the desperate need
to service the all-important self.

The US, UK and EU mantras of 'efficiency, transparency and

progress' are mirages created by a technocracy and bureaucracy that is as self-serving as its individuated subjects. It is a fragile unity that always breaks catastrophically when the money runs out. The EU is a rich man's family. Its members know no solidarity in poverty, only in wealth; and all their actions are ultimately directed towards the pursuit of wealth—even if at a superficial level they are convinced of their own good intentions. They adopt a worldview that states that it is obvious that it is our nature that society be based on the principle of 'each man for himself'. Everyone, except themselves, are true outsiders to them; even within their own tiny, secure enclaves. What model is this for human development? It is a model of inhuman development—one that we are expected to follow.

The unbridled arrogance with which the 'condition' of Africa and other marginalised regions is observed, analysed and expressed is revealing. The 'condition' is imposed from the outside. It is understood within a framework of development that is presumed to have already arrived—its technocratic overdependence is seen as a strength; the only appropriate response to 'vulnerability' (Bankoff, 2001). But this myth is faltering as the world changes at faster rates, and more discontinuously and surprisingly than ever before. Global change and climate change are testing this notion.

Katrina is a good example of this. Hurricane Katrina shocked America precisely because it was predictable and foreseeable, yet the devastation that resulted amongst poor, destitute communities that consist mainly of black citizens was reminiscent of floods all over the world. The current floods in Pakistan (2014) are far more devastating, but because Pakistan is not in America, nor in any first-world region, nothing about it is surprising. Indeed, it is viewed as normative. Indeed, these kinds of things tend to happen in those places all the time. But what of New Orleans then? What it revealed is precisely how mythical the idea of the 'first world' really is. It showed that despite the most advanced technologies in the world the political understanding and will of America was directed elsewhere—or more appropriately—misdirected. Many are correct in believing that if New Orleans was in Florida the cost in human lives and livelihoods would have been much lower.

This is not to say that there is no role for technology, but that there is no role for technology as it is historically and currently employed according to 'market economy' principles. Indeed technology offers hope for decoupling economic growth from resource exploitation and environmental impacts, but this will never happen as long as the democratisation of IP is prevented by the large military, industrial and political powerhouses of the world and their accompanying global 'lackey' institutions that reinforce their objectives within the wider world.

In addition, what of the inner cities of the US; where emergency food aid finds itself being distributed (Resnikoff, 2014), a cruel 'each man for himself' principle—itself a pathetic rehash of overly patriarchal, outdated, masculine notions of man as the hunter and provider—is unquestioningly espoused as synonymous with democratic principles. In my view, nothing could be further from the truth. The right to vote (or suffrage) is not an *a priori* condition or principle for maintaining an overarching belief that each person be set against each other as if they were animals in a jungle romantically living out the struggle for existence. Indeed, how can any functional society of human beings result from this? Self-interest, selfishness and dysfunction are direct results of the 'each-man-for-himself' myth that has created the current 'homo-economicus'. Indeed, whenever neo-liberal free-market principles have been broadly and sincerely adopted in 'transitional' and 'developing' democratic economies the very same social fragmentation and loss of community cohesion has quickly resulted, often accompanied by widening inequalities between small techno-savvy elites and the hitherto marginalised indigenous peoples or lower classes. South Africa is a good example of this. The Gini coefficient has increased drastically despite meeting the required level of economic growth which is touted as the solution to the very inequalities that bring about the sense of belatedness in relation to the West in the first place.

Neo-liberal laissez-faire (*'free market'*) principles have brought about precisely the opposite effects on the ground—no 'market equilibrium' is reached. Rather, transitional economies remain frail constructs; vulnerable to both external and internal developments. They remain 'metastable', balanced precariously on a thin edge. Strong feedback mechanisms need to be in place in order for it to remain stable so far away from equilibrium. Decision-makers are constantly playing catch-up while bounced around the "organisational white-waters" (Malhotra, 1999) in frail canoes. Unable to move up-river the only option left is to 'go with the flow' and hope that the final destination offers promise. We are no longer in a world where "All roads lead to Rome". We are in a world where all rivers lead underground. The river's course is set towards the subterranean sea of dissolution.

True masters of their own fate know that flows can be altered; they can be changed, redirected and rebirthed. So as the West diagnoses our condition as if it were an aberration of its own body, so too do we. We are complicit in our own distorted self-image. We see ourselves only through reflection in the mirror of the West. We stand, transfixed by our own image as if there are no other mirrors to explore—as if there are no other ways of seeing. We have assimilated both the mirror and the image precisely because of our sense of belatedness, our own inadequacy reflected in our notions of hierarchies of progress.

Trapped within the mirror we lose all agency, every moment reflected back remains trapped within a frame of reference that lies outside of our own experience and, hence, understanding. It remains mythical, as do we, trapped in its mercurial glare; unable to see ourselves for what we are, have been, and will be.

Yet can we question the 'inevitability' of modernity and negotiate it on our own terms? Is there a freedom that exists that allows us to achieve this? Indeed we can, but we need to immerse ourselves, to absorb vocabularies and cooperate with the multiple, and to see our own multiple projections and reflections within it.

Chapter 9

The Multiple in Action:
The Occupy Protests—Being Leaderless
Helps Reshape the Territory

At the time of writing this chapter, perhaps the most striking characteristics of the protests that have emerged over the past year and half, whether they have originated from the Arab Spring in the Middle East, or in the cities of the developed world, is that they are mostly leaderless, and that their various manifestations have proven to be devoid of a single governing ideology. Instead, they have become forums for participation, where a wide variety of views and ideological orientations can find a voice. As one Occupy Wall Street protester put it, "it is not so much an ideological movement as much as it is a space for participation".

In the Arab Spring, a necessary debate has emerged around how to reconcile Islamic governance with democratic principles, and the elections in Tunisia reflect this. There is also an emerging debate on the role of women in Islamic democracy. These are both welcome developments that attest to the power that the Arab Spring has brought to the general discourse on politics and political freedom in the Middle East, and the role of society and leaders within it. Islamic extremism has been put under the microscope and questions that were once thought unthinkable have now entered the mainstream. It's not perfect, but it's a beginning.

In developed countries where 'mature' democracies have been in place for the better part of the last forty years, the Occupy and anti-austerity protest movements have also re-opened the space for debate around the fundamentals of democracy. In the last 20 years, 'talk left, walk right' political strategies have emerged in the wake of the weakening of the left through the era of neo-liberal 'free'-market economics. At the same time, conservative ultra-right movements such as the 'tea-party' increasingly appropriate the language of the old left. In short, developed countries have become plagued by bipolar democracies where it has become increasingly meaningless whether a democrat/labour or republican/conservative government is voted into power, because they ultimately end up thinking and acting in the same way.

Whether one chooses right or left is an increasingly meaningless distinction in this muddled new territory of democracy, and it is obvious that the public responded with a profound apathy during the boom times. Now that the bubbles have burst, their voices are being raised. Those who were once content to sit in the middle and watch either side throw barbs at each other have entered the fray. They are not explicitly ideological, but they all want change. They may not all have the same ideas of how to make changes, but they know what needs changing. Simply put, it is unfairness at work in the heart of democracy. It is the violation of the egalitarian principle that lies at the heart of democracy itself. Inequality has always had strong appeal as a mobilising factor for those on the lower rungs of the divide, and when their ranks grow they become dangerous. It's as simple as that. And it's not perfect, but it is a beginning.

It is a beginning because these leaderless mass movements have opened up the space for participation, and participation constitutes the first step towards greater dialogue and exchange that can shape a new discourse. The 'leaderlessness' of these movements is precisely what makes them a potent force for change. That is, they are not already wrapped up in ideological clothing that predetermines how the debates and conversations may unfold. Instead, their 'leaderlessness' and lack of monolithic ideological posturing means that the opportunity now exists to host a social dialogue on what core principles are needed to navigate the territory of the new century, and what kind of changes might be necessary to realise these principles.

Perhaps the most significant demonstration that the protesters have given the world is that it is still possible to engage governments and institutions through direct protest action. Some may underestimate this, but for the past twenty years there has been a significant public reluctance to engage in direct protest action. This critically important regulating function of democracy gave way to a global wave of apathy, especially in the countries of the developed world. The triumphal march of a victorious capitalism brought with it a sense that capitalism itself was timeless, and the reigns that had previously kept markets in check were loosened, as markets themselves were seen as the 'self-regulating' magic-makers of prosperity. The results of this victory are now plain to see, and the hopes that things would 'self-correct' after the collapse are proving more distant than ever before.

We are now entering an era where new norms will be established. 'Business-as-usual' may find a few more tricks and turns to keep itself alive, but the *foundations* of business-as-usual have been shaken. The longer the recession persists, the more it becomes a rallying cry to the middle and working classes who have been displaced from real power in the oligarchical functions of the neo-liberal dream where there is 'one set of rules for us, and another set of rules for them'. The longer they remain leaderless in their protests, and refuse to be co-opted into simplistic sets of agreements that hijack their momentum, the more momentum will grow, ideas will flourish, and a new discourse and new normative orientations will be birthed.

Perhaps we are at a moment of the same significance as the modernists of the 20th century found themselves at, filled with the promise of a brave new world, yet unsure of exactly what it may end up looking like. Let's hope that this time we are able to avoid falling into the trap of conflict, and that our enhanced ability to share ideas, conduct campaigns and achieve genuine political action in different parts of the world can help us overcome the very real challenges we face. Should these leaderless protests become captured by populists who skilfully capture the 'vacuum' instead of celebrating the potential of the 'vacuum', then the risk of fragmentation and conflict will become more significant and perhaps overwhelm the momentum that, although currently loosely bound, may still yield a new set of ideas, visions and norms that effect change in ways that are as yet unforeseeable.

Again, true freedom is the ability to bring about fundamental change, so we might regard the 'leaderless vacuum' as a free space in which the birth of the new is possible. Straight-jacketing this space into a re-run of old positions and ideologies will prove to be its undoing. A parody of the ideological conflicts of the 20th century is hardly likely to achieve anything more than it already has, and in truth, these positions have already compromised themselves to each other, and to themselves, and have drifted far away from their original foundations.

And we do need the new to flourish, now more than ever. The Euro hangs on a thread, the dollar would be a worthless currency without China's strong backing. The 'free' market has failed because the most predictable of human behaviours, greed and the propensity to desire wealth without work, triumphed over the imaginary virtues of self-interest. We are now in trouble, and we have to work together to dig ourselves out of the mess we're in. An old supervisor, who oversaw me through probably the most difficult career transition I've ever attempted, once gave me some advice that has seen me through many difficulties. "How do you do it?" I implored. After some careful thought he replied: "I find that the most difficult thing is keeping the space open long enough for something to emerge".

In research, as in life, you often have no idea what you are going to discover or develop when a project begins; all you have is a bunch of ideas and speculations about how to move forward.

Yet if you keep the space open long enough, and stay faithful to the spirit of the project, something almost always emerges that makes it worth the wait. Perhaps the same kind of philosophical openness is required now, and hopefully we will find the strength to move into the unknown so that we can discover something better than what we have now, because more of the same isn't going to make the cut, and is more likely to sink the ship than to keep it afloat. Tossing the blame from one side to the other isn't going to solve anything, and we have to face up to the deep flaws that exist on both sides of the political divide, and soon.

Note: This chapter has been adapted from a blog post that was published online on 5 November 2011; see Peter (2011c).

Chapter 10

Freedom

The "lost treasure" of revolutions remains "nameless", fantastical (Arendt, 1961). So what are we left with to pursue, and to what extent can freedom be achieved and maintained? The freedom that represents the highest order of democratic equilibrium among participants is not just the freedom to participate, but the freedom to disrupt spaces in which freedom is curtailed and bring about change without destabilising the entire system—i.e. revolution without violence—the kind of 'revolution in perpetuity' that democracy potentially offers up a framework to support when it is sincerely, and not cynically, implemented as a political system.

"Whose world is this?" we may ask, as we reconstruct its boundaries through our interactions with the processes of information and communications technologies, the internet, globalisation, media, migration, mobility, urbanisation, education and economy at work in our everyday lives.

Perhaps we can escape into thoughtlessness as a way of evading the world. Perhaps we crave superfluity, especially when we are overwhelmed by events, thoughts and fears, and we escape reality by retreating to a more manageable form of it. Perhaps we just crave superfluity because it extends our experience of the world and momentarily enriches it.

It is, however, altogether natural for us, as social beings who construct our own reality, to construct realities that help us escape others. Even the realities of thoughtlessness, and of nonexistence, exists within the continuum of realities that we construct through "the multiple as such". That indeed is why it is multiple. Our reality contains all possible and probable realities.

Thoughtlessness is not excluded from the reality of the

multiple. Rather it pervades it. It is the most convenient retreat from reality. A loss of awareness, an immersion in sub-reality, provides temporary relief from reality, just as awareness provides perspective on it.

Freedom itself is not a thing. It is a word, a concept, an aspiration that cannot be achieved in any totality, except (possibly) with the advent of death. Likewise, revolution represents the *desire* for changes that can help bring about or facilitate greater degrees of 'freedom'. In my view, revolution is always mutation or trans-mutation in the service of a conceptual freedom that can never be attained. And so we must look to a more representative term for how change unfolds in society. For me, that term is 'evorevolution'. Part evolutionary, part revolutionary; societal change is reflexive. It is not deterministic, and cannot ever realise a complete break with the past. It can declare a history, but it cannot outrun the past, which manifests in the various interactions and influences that combine to yield the changes that unfold within the social fabric of the multiple.

The reality of the multiple is that both the processes of consolidation and change are simultaneously at work, albeit to different degrees in different contexts. Revolution and all attempts to 'reset' the initial conditions of a society so that a new order can emerge are hence always condemned to failure. Intervention and disruption are the real vectors through which purposive change is directed within the multiple.

The process of change, however, is contingent on the properties of the multiple. The multiple is heterarchical, uncertain, undecidable at times, self-organised, emergent (both in the perceptual and material sense), richly connected and profoundly open. It is complex, and its complexity is embedded in the processes of change. Change cannot escape the irreducibility of the multiple. It is always, in part, a product of it.

Chapter 11

Evil and the Multiple

According to Hannah Arendt (1971), it is thoughtlessness that generates evil, that is, the "absence of thought". The antecedent for evil is the inability to think from another perspective or to 'put oneself in another's shoes', so to speak. The absence of thought as a condition for evil means that evil is not radical but banal—*an everyday evil*—that allows good, ordinary 'family men' to commit extraordinary evils. It is a particularly keen insight into how extraordinary evils are upheld within a greater social fabric, i.e. without question. Indeed, Apartheid society was not radically evil itself; it persisted precisely due to how well Apartheid ideology became spoken as an everyday truth. Having lived under Apartheid, it is difficult not to applaud Arendt's deep and compassionate insight into the nature of evil. At the time that she shared her view with the world she was roundly attacked for the 'coldness' of her analysis. And while her analysis may have been objective—in as far as that was possible in the case of Otto Adolf Eichmann—I would argue that it is profoundly compassionate in its implications (indeed, as she herself does). When we accept the banality of evil, only then can we espouse the most noble of virtues: forgiveness. When we recognise the banality of evil, it enables us to proclaim, "Forgive them, for they know not what they do!" with sincerity. It forces us to recognise the humanity of both victim and perpetrator, and especially the flawed nature of human existence. We have never been, and will never be, perfect.

Yet the origin of this insight—that the "absence of thought" is the source of evil—begs further enquiry. Adopting the perspective of the multiple (i.e. within and without Lazarus) to understand this absence of thought leads to an interesting analogy. The "absence of thought" can then be likened to the vacuum. It is not empty. The multiple helps explain the substance of the vacuum as it manifests in Lazarus' mind. The 'multiple as filter' hence has great relevance for understanding Lazarus.

"Thoughtlessness", in this sense, is not so much a vacuum as a noise. The more noise, the less thought is put into action. However, the converse is not necessarily true; that is, there is no corollary! Moreover, non-linearities may subvert both forward and converse relations. Noise has the potential to destabilise only because it has the potential to overwhelm.

Evil, viewed in this way, cuts across classes, genders, races, creeds, sects, elites and so forth. It *is* present in people, but manifests or emerges as a product of a range of factors that lead to thoughtlessness. Its origins are hence tricky to establish fully, precisely because it emerges from a dull noise, and not a clear signal. Hence it can never be regarded as truly radical. Anything truly radical is accompanied by a clear signal that distinguishes itself from noise. Only when a radical belief becomes banal and 'everyday' does it become truly evil, because then it can manifest with the aid of the multiple.

So what then of precedent? How is precedent established? Who decides "never again!"? Who enforces it, and how? All this depends critically on how we conceive of the way evil manifests in society, and from what origins it hails. Casting evil as radical or banal profoundly affects how we attempt to deal with it, and what success we achieve in realising "never again" aspirations, which to date have failed humanity over and again.

A break with the past is impossible. We have to live out the future with an awareness of the past.

Futurology must be sensitive to what interventions and disruptions are emerging, or can potentially emerge, even those that may seem to lie outside of the probable. At the same time, the futurologist must be acutely aware of how the past manifests in the present, and may manifest in the future, i.e. to understand what cycles are in play, hidden within the memory of the multiple.

To establish precedent requires more than constitutionality and policy. It requires a shift; where intervention(s) and disruption(s) establish new trajectories within the multiple that become normative in the *experience* of the multiple.

So when will the egg crack? When will the new South Africa actually be born? And when will the new Africa be born? The incubation has been too long. Something has to be birthed.

Apartheid attempted to create a break with history; to *engineer* a whole new social reality and to make history subservient to it instead. As a consequence, whole histories were subordinated to the national narrative of "separate development", whether out of shame or expediency. The multiplicity of identities became allowable only within race categories, but still, some held on to aspects of their multiplicity. However, for every race group in South Africa, the obliteration of memory beyond three to four generations is all too apparent. It was necessary for the Apartheid project. And now, what remains of the distant past remains hidden in the fabric of the multiple. It has not been raised up, assembled into a coherent narrative that we can call history. But without this history, all we are left with its various manifestations—often cyclical—that arrest us in moments that we cannot understand, where we have no context within which to assimilate/interpret them.

All determinist, positivist systems (whether Marxism with its analysis of economic history and projection of a final class struggle; or neo-liberal capitalism with its 'end of history' projections) become crude instruments when deployed in the multiple. They exert a 'violence' on the complexity of reality by attempting to reduce it sufficiently so as to be easily understood and propagated, hence ensuring the survival of ideology. They thus become instruments of control and power, rather than theories that adequately represent social reality. This reality—"La Belle Noiseuse"—is unavoidable, and without intervention inevitably 'strikes back' against these clumsy attempts to represent it. All that is left, then, is *faith*.

Marxism sets up its argument by arguing on the basis of socio-economic history. However, both Marxism and indeed even Thomas Piketty's argument are vulnerable precisely because of their heavy reliance on projecting the future on the basis of the past. That is, neither caters for the possible, only the probable. Yet the multiple is complex. It hosts emergence. It hosts spaces of possibility alongside probability, and so cannot be easily captured by a set of logics that linearise the interplay between past and future.

Thomas Piketty (2014) has argued convincingly that economic growth, when accompanied by higher returns on capital, is historically linked to increases in levels of socio-economic inequality. Since antiquity, global economic growth has only exceeded returns on capital in the period spanning the late 19th century into the first half of the 20th century (i.e. around 1950), and has since reversed again. Hence Piketty warns that the prognosis for the future—in terms of inequality—is potentially dire. Yet framing the arguments in terms of strictly linear relations between past and future prevents us from exploring possibilities in which greater equality might still be feasible despite high levels of wealth accumulation. It negates the potential for innovation that we, as societies, hold within our imaginations and our ability to self-organise. Thus, the agency of and within the multiple is denied. It is as though we cannot exercise our influence on the multiple to bring about new trajectories. And so, all that is left is for these new trajectories to be discovered by accident; that is, if they are recognised at all.

It is not difficult to imagine a mind to which the sequence of things happens not in space but only in time like the sequence of notes in

music. For such a mind such conception of reality is akin to the musical reality in which Pythagorean geometry can have no meaning.
—*Rabindranath Tagore: In conversation with Albert Einstein (Gosling, 2007)*

Indeed, Hannah Arendt was critical of Kafka's "rectilinear" understanding of time. In "Between Past and Future" (1961) she observes that the most recent history is still being lived in the *gap* between past and future. This gap only exists due to man's presence and the fight he maintains against the forces exerted by past and future. Arendt writes of a "diagonal line" through history as the way out, which would allow the "fighting" man to rise above his eternal battle against past and future. I believe that this diagonal line is better articulated through the notions of intervention and disruption, as presented later in Chapter 13, "Intervention and Disruption: Tools for Change" in Part II of this book. That is, when the filter is the multiple, it allows for non-linear understanding to emerge.

Arendt had a profound understanding of how people and their psychological, social, political and economic orientations and roles in society generate socio-political change. She ultimately came to view radical evil as an impossibility, because it implied that evil possessed deep roots. In her judgement, evil may manifest as extreme, but it is always a product of the banal and superficial; it could "lay waste the whole world precisely because it spreads like a fungus on the surface" (Arendt, 1978). She understood freedom as inextricably linked to action in the public realm ("… for to *be* free and to act are the same"; Arendt, 1961), and conceived of power as the potential that resulted from the ability to "act in concert" (Arendt, 1958). As she put it, "historical processes are created and constantly interrupted by human initiative" (Arendt, 1961).

In the perspective adopted in this chapter, rituals and routines maintain thoughtlessness. Awareness, however, brings about the possibility of change. Mutual awareness brings about the ability to "act in concert" and to yield the freedom that accompanies possibility. By acting together, we may be able to reclaim the "lost treasure" of revolution. We only become doomed to repeat the past when we are unaware of the non-linearity of time.

Part II

The Multiple and the Future

Chapter 12

The Multiple as Filter

Truth, which is one with the Universal Being, must essentially be human, otherwise whatever we individuals realize as true can never be called truth—at least the Truth which is described as scientific and which only can be reached through the process of logic, in other words, by an organ of thoughts which is human. According to Indian Philosophy there is Brahman, the absolute Truth, which cannot be conceived by the isolation of the individual mind or described by words but can only be realized by completely merging the individual in its infinity. But such a Truth cannot belong to Science. The nature of Truth which we are discussing is an appearance—that is to say, what appears to be true to the human mind and therefore is human, and may be called Maya or illusion.
—Rabindranath Tagore: In conversation with Albert Einstein (Gosling, 2007)

So to the question of the multiple as filter (or, indeed, multi-filter), which is the subject of this book. What (multi-)filter can serve the complexity of the multiple? For this, one has to look to the properties of the multiple, in all its complex glory. And note, it is not the mechanisms of the multiple that one must look to, but its *properties*; the properties that ascribe its complexity.

Discussion of Complex Systems Properties
Emergence:
The first of these properties is *emergence*. Emergence refers to behaviours that arise unexpectedly, and which could not have been predicted using the historical behaviours of a system or assemblage. Emergence may arise through two different instances.

In the first instance, emergence is understood as emerging from the complex, non-linear interactions within the richly connected, dense fabric of the multiple (Baas and Emmeche, 1997). This is how emergence is most commonly understood. In this instance, behaviours do not emerge as they are expected to from a model that draws on historical observations of the system or assemblage to predict or project its future behaviours. This can be referred to as *ontological* (or epiphenomenal) emergence.

In the second instance, and less commonly understood, emergence results from the multiplicity of perceptions on—and from within—the multiple itself. That is, we all recognise the 'whole' differently because we all 'see' the multiple differently i.e. partially. None of us can 'see' the whole in its entirety. Its multiplicity is too vast. Moreover, the notion of the 'whole' is created and agreed upon by us in the first place. Except in the metaphysical sense—i.e. in joining with the multiple absolutely—we cannot know the entirety of the multiple. This is referred to as *perceptual* emergence (Islam et al., 2006).

That is, emergence is defined in terms of a duality. It is both about the finer interactions of all 'things' within the multiple, as well as about how we perceive it from within it. While the behaviours of the multiple may result from interactions within its fabric, it is also co-produced by us, and is framed by how we perceive it. That is, we generate the emergence of the multiple from within it through perception, even as it is generated from material interactions within it. Its emergence is a product of this duality.

Self-Organisation:
And as it is with emergence, so it is with self-organisation. Self-organisation also has two faces. In the first instance, self-organisation is contingent on the adaptive capacity of a system or assemblage. In turn, this itself is contingent on how functions, controls, processes and networks can reorganise themselves to

retain themselves within a particular state, or to transition to another.

In the second instance, as with emergence, self-organisation is a direct product of the filters or 'lenses' that are used to observe the system or assemblage. An interesting study revealed how the entropy of a system can be different for the same system, depending on which lenses or filters are applied to it (Gershenson and Heylighen, 2003). Entropy reveals the extent of organisation of a system. The higher the entropy of a system, the less organised it is. If everything in the multiple that constitutes the Universe tended towards perfect entropy, everything within it would be dust. There would be no structure, no form. For structure and form to appear, we require uncertainty, chance and randomness to exist as key properties of the multiple, for they generate structure and agglomeration.

And so, self-organisation—as a property—is also constituted as duality, alongside emergence. Any serious undertaking to embrace the complexities of emergence and self-organisation must hence necessarily adopt this duality as central to the properties of complex systems and/or assemblages.

Heterarchy:
A heterarchy is an evolving, changing hierarchy. That is, it is an impermanent hierarchy. In a sense, it is an open hierarchy. On the one hand, uncertainty, chance and randomness govern the evolution of the heterarchy. On the other, self-organisation within the heterarchy also governs how it changes and evolves.

The multiple is heterarchical, primarily because it is open and constituted by rich densities of relationships, some of which have higher and lower probabilities of activating under differing system assemblages and/or sets of circumstances. Heterarchy can be envisaged as a "fishnet" in which controls, processes and functions "rise to authority" within the fishnet, constituting, in a sense, temporary hierarchy of governance of and/or within the multiple (Heylighen, Joslyn and Turchin, 2001).

But this two-dimensional perspective is not the only way of thinking about hierarchy. We can also think about hierarchy in multi-dimensional assemblages. In the case of multidimensional systems and/or assemblages, hierarchy is more than just where controls are located within the system. It is what ascribes the recognisable form that a system takes in respect of its multiple dimensions. It gains its shape and form within the multiple through the hierarchy that ascribes it within the multiple. Yet at the same time the shape and form are also subject to perception, and preconception. We within the multiple obtain agreement, but to varying degrees, on what constitutes the hierarchy of the multiple as we journey through it. We construct the hierarchies that enable recognition of the identity of the whole, and the identity it acquires. *We make social reality, as much as social reality makes us!*

Degeneracy:

At the same time, the behaviours that emerge at a whole-systems level may also be generated by *different* assemblage configurations and conditions. Degeneracy refers to systems where different sets of configurations and conditions yield the same overall system behaviours. It is a useful phenomenon (i.e. multiple stable states), because it allows for systems to retain their same overall 'identity'—i.e. in terms of observable behaviours—by being able to exercise controls in order to fall back into different system configurations and conditions in response to exogenous and endogenous changes and pressures. The more 'degeneracy' a system or assemblage has, the more it is able to retain its identity and boundary conditions in the face of exogenous and endogenous drivers of change. In short, it becomes more *resilient* due to its degeneracy.

To acknowledge degeneracy is to acknowledge the multiple; in particular, it's fundamental heterogeneity. It is this heterogeneity, and plurality, that enables the multiple to adapt and adjust to retain a particular *identity*, even though it may be generated by different sets or configurations of structure, controls, processes and networks. This identity is itself a duality, as it is both generated by the ability of the multiple to adopt multiple stable states (or self-organise into different "basins of attraction"), as well as how we perceive it to be self-organised.

Sub-Optimisation:

And linked to the concept of degeneracy is that of the "sub-optimisation principle" (Skyttner, 2001; Richardson, 2005). If the overall system is to be able to functional optimally, then its sub-systems cannot all function optimally. The converse is also true. In a sense, this suggests that trade-offs are required between sub-systems in order to achieve the balance that is necessary for whole-system optimality. In a simple sense, this points towards the need for broader inclusion in critical decision-making. Note that degeneracy enables sub-optimisation, or opens up potentials for sub-optimisation, and that ensuring system redundancies is critical in this regard. That is, simply optimising a system or assemblage is not enough to guarantee its ability to adapt and evolve. It requires footholds, options within it, to be able to adopt different strategies when encountering uncertainty and risk, and to harness its adaptive capacity to remain resilient in the face of change.

Undecidability:

And it is with particular reference to the need for negotiating sub-optimality between occupants of the multiple and their various interests that the notion of undecidability becomes important. The undecidable is Plato's "included middle". In theory, it refers to irreconcilable metaphysical opposites. For example, a decision may be viewed in strict metaphysical opposites such as "good versus bad", "evil versus virtuous", or "fair and unfair" (e.g. in the case of trade-offs). According to Derrida (1992):

> *The undecidable is not merely the oscillation or the tension between two decisions, it is the experience of that which, though heterogeneous, foreign to the order of the calculable and the rule, is still obliged – it is of obligation that we must speak – to give itself up to the impossible decision, while taking account of law and rules. A decision that didn't go through the ordeal of the undecidable would not be a free decision, it would only be the programmable application or unfolding of a calculable process. (In Banisch, 2002)*

It is not difficult to envisage how easily undecidability manifests. It is, in reality, encountered daily, and lends complexity to everyday life. It governs the uncertainty, contestation, conflict and dispute that arises when decision-making is required in complex environments, or when facing complex challenges. For example, when two different sectors compete over the same resource, and are limited by it, complex negotiations and mediations are generally required in order to reach agreement over the way forward.

Some undecidable decisions may not be fully resolvable to the satisfaction of all parties. Hence it is important to facilitate these negotiations and mediations through engendering a strong *shared understanding* of what makes a particular decision undecidable. This is the foundation from which undecidability must be encountered. And it is from this shared understanding that trade-offs and perhaps even sacrifices can emerge.

Antifragility:
Antifragility (Taleb, 2012) refers to the ability to harness adversity and difficulty, and acquire an advantageous disposition from encountering it. While degeneracy allows for a system or assemblage to retain its identity (i.e. its adaptive capacity), it does

not account for the ability of it to transition or transform to a wholly new identity, replete with different structure, controls, processes, functions and intra- and inter-dependencies. In order for this to occur, the ability to innovate is required. This in turn is dependent on *creative capacity* (note: *not* adaptive capacity). In this respect, the ability to be able to envisage and/or identify *new* possibilities and potentials, evaluate them, test them and refine them is necessary. In a broader sense, this emphasises the need for niches to exist and thrive within the multiple. And in addition to creative capacity (which resides in the imagination, different ways of seeing, perceiving and interpreting), hetero-geneity and redundancies within the multiple are necessary for antifragility to manifest as a key property of the system or assemblage in question.

And it is important to note that redundancy and optimality work in opposition to each other. All efforts to optimise invoke efficiency criteria, which require reducing redundancies in a system or assemblage. And in this way, highly optimised systems actually become fragile. When configurations of structure, controls and processes that make up the system or assemblage become highly optimised, they also become more rigid (even though well capitalised, i.e. in relation to any resource). It is this rigidity that makes it vulnerable to exogenous changes that lie outside of it, which, when manifested, render the highly optimised system incapable of adapting its structure, controls and processes to either transition into a degenerate state, or transform to a wholly new state entirely while retaining its purposiveness.

Uncertainty:
Uncertainty is latent in the fabric of all interactions within a system or assemblage that resides within the multiple. While uncertainty can be associated with variability of sub-components within the multiple (in which case it can be regarded as a

mechanism), uncertainties can also accumulate and manifest at the 'whole' system or assemblage scale to produce variation. And yet again, how this whole system variation is perceived—i.e. according to what lenses or filters are used to observe it—yields yet more uncertainty, i.e. a perceptual uncertainty (in the same way as emergence and self-organisation can be regarded as perceptual). And so, in the case of uncertainty, it also constitutes a duality, and not a dualism, within the multiple.

Non-linearity:
And as it is with uncertainty, so it is with non-linearity. It is both a property of the whole, as well as a mechanism within the whole. The whole system or assemblage can exhibit non-linearity—i.e. in its observed behaviour—as a consequence of the ability to transform itself into an entirely different modality or state; whether through self-organisation, self-disorganisation or due to overwhelming exogenous pressures that exert large pressures on it, forcing it to transform.

Conclusion:
The properties of complex systems or assemblages that reside within the multiple that have been articulated in the previous sub-sections do not constitute an exhaustive set of properties, and are in fact themselves subject to a particular 'way of seeing' the multiple. As such, this does not constitute a closed, grand theory of the multiple, but rather an open-ended theoretical framework that can evolve and change over time itself. Only a theory that itself has the properties of the multiple can possibly and/or reasonably attempt to understand it. And so, it is with this in mind that this theoretical framework has been formulated as the "multi-filter".

It is a non-reductionist perspective. It is also open to the possible, the new, alongside the historical. It does not discriminate between them, but regards them as all part of the continuum of the multiple. It opens up room for exposing and interrogating complexities. It also enables simplification where necessary, but does not simplify before embracing complexity. Rather, it forces direct encounter with the complex before arriving at simplicity, if it is indeed a possibility at all. This is important, as over-simplification at the outset of any analysis renders what is being analysed to a mere remnant—or abstraction—of itself, and false diagnoses and prognoses most likely result from that approach. In other words, utilising a simple implement may falsely render the task to a simplicity that does not reflect the complexity of the task. That old adage "when all you have is a hammer everything starts to look like a nail" becomes relevant here.

And there is more to deploying filters—which are derived from an understanding of the complexity of the multiple—upon it. It may seem perverse, a cannibalisation at work, but it has precedent. For example, more recent developments in particle physics have shown that in order not to violate the uncertainty principle—where obtaining a very precise measurement of one variable (e.g. speed) of a particle requires a sacrifice in knowledge of another variable (e.g. position), and vice-versa— the filter of measurement itself must be made 'fuzzy' or less intrusive. And it is in this spirit that identifying the *properties* of complex systems and assemblages within the multiple, rather than its *mechanisms*, becomes an important departure point for developing a 'way of seeing' the multiple. In this framework, which begins from a consideration of the aforementioned properties, whatever mechanisms are relevant are an outcome of the analysis, not the starting point.

The framework—or multi-filter—does not promise clear answers. It only enables the user—Lazarus—to investigate, probe, and understand the multiple, and answers are left to the user to determine; as an outcome of the understanding gained through embracing a multi-perspective or multi-dimensional analysis. The framework is a framework of tools for attempting a journey through the forested jungle that is the multiple, and all are necessary for good navigation towards a destination. They are not, in and of themselves, a route to the destination. That route must be discovered.

But more importantly, the multiple, when approached this way, ceases to be an object of mere ideology, or positivist inquiry. Rather, the multi-filter reveals the various dimensions and faces of the multiple, forcing the analyst—Lazarus—to engage with it head on. Lazarus, within the multiple, hence becomes empowered within it, able to relate to its various dimensions in a more practical manner. It is a way of navigating the noise of the multiple, and finding and exposing the many, diverse signals within the noise.

The multi-filter does not invoke the traditional scientific method as its starting point. Induction and deduction are secondary in this framework. Rather, this framework deploys abduction as the entry point. That is, the abduction of a concept or *metaphor*—i.e. such as the properties identified in this chapter—from one problem context, to its application in another context, which allows for new patterns and causalities to be identified and recognised.

Charles Peirce (1934) observed that abduction, and not induction and deduction, were responsible for the great leaps forward in science. And when encountering the complexity of the multiple, we would only embark on a project that leads to sheer madness or at the least utter confusion, should we seek to engage the multiple through induction and deduction. That is not to say they are irrelevant, but rather that they are secondary—in the sense that they are complementary and not central—to abduction in this particular framework.

This framework does not prescribe a new way of engaging the social, political and cultural. Rather it provides a framework for unlocking new ways forward. Its power is that it does not begin solely from ideology, historical evidence or praxis as starting points. Rather, the *properties* — as outlined here — can be said to constitute a higher level of abstraction than a set of *principles* for engaging with the complexity of the multiple, i.e. on its own terms, rather than ours. This negates the invocation of morality, for example, as a starting point for analysis — e.g. dividing things into good and bad — and helps Lazarus avoid ending up in the inevitable trap of moral relativism — or ideological hegemony — that emerges when that kind of approach is adopted as a starting point for engaging with the complexity of the multiple.

This is a framework that allows Lazarus to immerse himself in the multiple, to swim in it, breathing here and holding his breath there, rising above the water and plunging beneath it, traversing the great expanse of the sea with a set of instruments, a compass, a knowledge of the stars and trade winds, a telescope and a sextant. This is a framework that empowers Lazarus to traverse through the multiple when he has no map, and is perhaps not even sure of what his eventual destination is. And to be sure, it is not the only framework that exists for this purpose, for it is one amongst the multiple, and that is how it should be.

It is not only knowledge of these principles and properties that is required, but an awareness of how they manifest in the multiple is important to acquire. Because the framework is abductive — i.e. dependent on metaphor (e.g. heterarchy, self-organisation, undecidability, sub-optimisation principle) — its application does not involve "the mere unfolding of a calculable process". It necessitates engagement with the multiple; seeing it through new eyes and filters over and over again, focussing here while de-focussing there, then resolving here while un-resolving there. It necessitates not only observing and measuring, but intervening and adapting. It requires not only the construction and formulation of models but their constant interrogation and deconstruction at the same time. The aim is not to construct a perfect theory, but to progressively internalise these principles through engaging with multiple contexts that reside within the multiple. Through this, I argue, a new understanding of social reality can be obtained. An awareness that becomes internalised, and mirrors the properties of the multiple.

Chapter 13

Intervention & Disruption: Tools for Change

A disruption is a non-linear intervention in that it forces a rapid shift or transition of a set of flows from one distinct state to another. Intervention and disruption are stocks in trade for leaders, strategists and philosophers alike. Intervention and disruption are tools that enable us to formulate strategies for change. They cut across time, scale and sector. They achieve this by stopping existing flows and cycles and establishing the basis for new cycles and flows to emerge. This, in turn, may or may not lead to changes in purposiveness as well as identity.

Which flows are disrupted and intervened upon? The flows that are disrupted are the flows that maintain hierarchies of values, beliefs, norms, behaviours, narratives, discourses, definitions and classifications (or taxonomies), causality and cyclical patterns or processes. Information, influences, materials, and decisions that flow through the social fabric of the multiple are targets for disruption. So is the sense of space, time, individuation, inter-relation and identity. In this framework, a regenerative strategy would attempt to disrupt cyclical patterns that have become frozen, or fixed into destructive flows.

Finding counterpoints in the rhythm of flows enables one to disrupt flows and 'flip the switch' — in a sense — from one state to another, with minimal effort and intervention. Counterpoint is found by understanding or sensing the conceptual framework that is at work in establishing the rhythm; so it can be destabilised.

A design without a strategy is a design void of concept; it does not 'feel' the rhythm. It is only an idea and has no conceptual framework in which it is located. A concept can disrupt and intervene upon the status quo by making different connections evident, revealing a new understanding, previously hidden by the filters or lenses in use.

A concept is different to an idea. A concept establishes a framework in which many ideas can reside, but an idea alone does not constitute a concept. An idea is a disconnected island while a concept establishes a field of ideas that can be played out in relation to each other in respect of the said conceptual framework. True strategy is conceptual. It abstracts the strategic essence so that an adaptive framework of ideas can be 'matrixed' together in different combinations to meet varying scenarios appropriately.

Intervention and disruption are *heterarchical*. They can act across and subvert hierarchies, and therein lies their great power.

Whereas the multiple as filter is about understanding and diagnosing within the multiple, intervention and disruption are about *action*. Developing a sensitivity to flows and rhythms is essential for action, as the timing of actions is in part critical to the success of an action or a set of actions. As previously argued[4], it is not enough to intervene without a sense of rhythm and flow, for without it there can be no sense of counterpoint. That is why it is not enough to merely measure flows, or to diagnose the multiple using its properties; it is important to internalise them in relation to the multiple.

It is the sensitivity to rhythms and flows that enables the capacity to take action through intervening and disrupting. Intervention and disruption, in turn, gives rise to the new (or at the least potential for the new), but the new—i.e. Hannah Arendt's "natality" (1961)—cannot be controlled. It is always part design, part emergent, and takes on a life of its own within the multiple. And so, despite the greatest fantasies of those who dream of being able to engineer social reality, such efforts always end in frustration at best, or authoritarianism, totalitarianism or some combination thereof at worst.

But which flows and rhythms am I referring to here?

Flows, for example, could be constituted by influences (e.g. power), materials, information or decisions (e.g. command and control). Which flows are observed or identified depends on the filters of observation that are employed.

Rhythms are ascribed by cadence and regularity in the movement of flows. Rhythms can be multiple, and can hence manifest as noise when unfiltered. And so, rhythm also depends intimately on the filters that are used to observe or identify it.

Both flows and rhythms result from the self-organisation of systems or assemblages within the multiple. They indicate that something is bounded within the multiple and has taken on shape and form within it, has broken from the noise and assumed a dynamic structure. It begins to pulsate and breathe, and so becomes 'living' in a sense; recognisable, name-able, something that can be experienced.

But there is more to the 'filters' that observe rhythm and flow. The senses—or the "dance of the senses" (Serres, 1982)—bring rhythm and flow to life. That is, there is something inexplicable about good timing, about choosing a moment, about knowing when to act. Sometimes, of course, it is just pure luck. And this simple fact always reminds us that choosing the right moment remains a game of chance, and it is in that spirit that action must be taken, for certainty is only the luxury of the dead, not the living.

Intervention and disruption have many vehicles and instruments. Niches that reside within the multiple, for example, can accumulate and grow, eventually disrupting the regularity and normative basis of systems and assemblages—i.e. bounded entities—within the multiple. Sometimes intervention and disruption can be subtly induced, through a series of "nudges"; for example, defaults that set in motion a set of subtle influences that guide change, rather than inducing or *forcing* it (Thaler and Sunstein, 2008). Yet brute force can also be deployed as an instrument for intervening and disrupting. It is an active force. In contrast, obstruction is a passive force. And so *force* is the underlying instrument in both the aforementioned.

Intervention and disruption can also be driven by changes that occur within and between agents in a system or assemblage. Changes in the values, beliefs and norms that govern their behaviours may lead to the emergence of new behaviours, new limits and new thresholds. Whether individuated or networked, changes in values, beliefs and norms can bring about significant radical *or* incremental change.

Purposive change, however, requires further deliberation. It necessitates both subtly influencing factors—thereby helping to seed and catalyse change—as well as the application of force when necessary. Drawing on the metaphor of a fighter, the key strategic elements that combine towards a particular desired end are; *positioning, orientation* and *awareness.*

'Position' refers to the location of the actor (or actors) within the multiple, in relation to the subject of action (i.e. a system, assemblage, person, etc.) and from which an intervention or disruption is launched. Without good positioning there is no stable ground from which to act and actions are made superfluously, in the vain hope that they will influence the subject in the way that is desired. The ability to change position—and to make small adjustments in doing so—in order to gain an advantage or measure of influence over the subject, even catch it by surprise, is what positioning is about. An intelligent, intuitive knack for positioning can put an opponent off-balance, making it easier to disrupt their flows and behaviours. As it is with a game of chess, you can only work from the position you have, or the position you gain through positioning.

In turn, orientation is about what perspective is held on the desired change. There are many different orientations that can be held on how to achieve the same objective. In that sense, orientations can exhibit degeneracy. They can be complementary, antagonistic or agonistic when compared with one another, but they all essentially pull towards the same objective. In some cases, orientation is simply preference, but in others it is a consequence of a wealth of knowledge and experience. Orientation sets the direction for intervention and disruption. It does not guarantee the direction, but it sets it.

Awareness is the particular sensitivity—in the context of disruption and intervention—that enables one to choose the moment of action by being sensitive to the contextual specificities and dynamics in play. Awareness is necessary to feel the 'rhythm' and to choose the moment when an action is taken. Like a dancer, a fighter or a jazz musician, awareness—born out of long hours of training and being in the space of action—in choosing the moment of intervention requires exceptional intuition and timing. Awareness is not merely calculated, it is felt!

And so, when looking to the metaphor of a fighter to negotiate complexity, position, orientation and awareness become more important than acting on perfect or near-perfect knowledge and plans. It becomes more important to train intuition and instinct. They are the critical foundations through which knowledge is filtered and actions are premised. There is no perfect knowledge—no perfect models—upon which movement into the future can be based.

And yet there is something more about intervention and disruption that cannot be left unsaid. It requires that the *will* be exercised. As Hannah Arendt (1971) states: "… because the Will, if it exists at all … is as obviously our mental organ for the future as memory is our mental organ for the past ".

And so we return to the fundamentals of what constitutes the past and the future, will and memory. It is not dichotomy. The will cannot be exercised independently of memory, and must remain sensitive to it, lest it lose its way and create conditions to relive the past instead of ushering in the new. And so, without the will, and without memory, all intervention and disruption, in essence, ceases to be purposive. And yet, even when intervention and disruption are purposive, their outcome cannot be guaranteed within the heterogeneity of the multiple. In short, many outcomes are always possible. Shaping them is the task of the leader, while understanding them is the task of the philosopher.

Chapter 14

On the 'Ethical Idea'

Anyone who works long enough with their tools comes to terms with their inability to reduce or capture the reality of what they are dealing with. Moreover, our tools become invested with us and so we employ them as extensions of ourselves. They reach the object, so it is always part subject, when caught in the filters of the tool. When we reach into the problem spaces of the world with our tools, multi-modal and multivariate—in search of the whole—we distort interpretation through the act of intervening[5]. So intervention does not always have to be direct. It can be directed strategically. For example, in order to score a point in sport one has to create openings and seize them, and be responsive when they emerge; it is often insufficient to follow the procedures of strategy, even perfectly. That is, a strategy should open up spaces of opportunity, like a boxer, feinting and adjusting so as to open up opportunities in the future—not aiming direct, singular blows. A discourse is subject to the same limitations, and is worthless without the disruptions and subversions that are made to open new spaces in the discourse. Improvisation and doing something unexpected can change the game, but it is a double-edged sword to those desiring control in the absolute, who cannot appreciate opportunity as anything other than 'risk', and so frame the interventions as conservative, risk-reducing strategies that close, rather than open, the space for opportunity and innovation; more concerned with controlling society rather than facilitating society.

So how do we contemplate the impossible, the irreducible and uncertain? Do we render them banal, or do we exalt them for our own purpose? Undoubtedly both. For example, religion is proof of how all three factors are negotiated in everyday life. In fact, religion makes currency of these factors in administering principles for everyday life, but increasingly, as religion recedes in society, and with nothing to replace it, these factors are rendered to the realm of the banal. They are remade into terms for everyday consumption and regurgitated without meaning. In the arena of entertainment, of sport and the sexual, for example, uncertainty, anticipation and surprise titillates, antagonises and draws us into the spectacle of drama. We are seduced, transfixed by uncertainty in banal spaces. Indeed as a consequence there is no appreciation of the impossible or the realm of the possible that it gives birth to, which some argue comprises everything of significance; the variation that brings life, mutation and the new. Rendered banal, we barely see what is possible, are blind to the impossible, and the future is consigned to mediocrity. We remain unconscious of what is emerging around us and cannot fully participate in our own future. According to Mahatma Gandhi (1922):

> It is the moral nature of man by which he rises to good and noble thoughts. The different sciences show us the world as it is. Ethics tells us what it ought to be. It enables man to know how he should act. Man has two windows to his mind: through one he can see his own self as it is; though the other, he can see what it ought to be. It is our task to analyse and explore the body, the brain and the mind of man separately; but if we stop here, we derive no benefit despite our scientific knowledge. It is necessary to know about the evil effects of injustice, wickedness, vanity and the like, and the disaster they spell where the three are found together. And mere knowledge is not enough, it should be followed by appropriate **actions**. An ethical idea is like an architect's plan. The plan shows how the building should be constructed; but it becomes useless if the building is not raised accordingly. [Emphasis added]

This study differs in that while there is agreement on the dual nature of the 'windows' of the human mind, there is disagreement on the metaphor of a 'building'. In the planning of a building, the plan is revised many times over in the mind of the architect before it is finalised. After that, the 'building' is a mostly rigid construct, which can only change with wear and tear over the long term, or radical change in the short term (such as breaking it down and rebuilding it). Once set in place, changes can only be made in relation to the design or plan. However, the ethical idea as viewed in this book is a house that is being deconstructed as it is being constructed. That is, these are two indivisible processes that must be in place for pluralism and diversity to be maintained within an ethical idea. The ethical idea is no longer a lofty moral, or even principle, which itself can be 'impossible' in its conception but is a space for contestation and cooperation around how we see ourselves and the world we live in. It is more akin to a game of meaning.

Chapter 15

What Kind of Theory?

What is theory? It is abstracting to higher levels of interpretation—replete with concepts and conceptual frameworks—to
distil, as far as possible, the essence of the subject, and/or the
properties that ascribe the subject, and the principles that govern
the subject and the theory of it.

The quest for the subjects' essence consists in determining the
set of attributes that its identity and function are premised on.
The essence consists of *what* is unique to a particular subject,
whatever it may be.

In the interpretation of what constitutes a theory for complex
systems or assemblages proposed here, the properties of complex
systems are not what make the system work (those are mechanisms). Rather, systems properties govern or describe their
behaviour and identity.

So this theory is about what the properties are that
govern/describe the subject and the theory of it, and how they
manifest. Properties, in this framing, are higher levels of
abstraction than principles (e.g. the sub-optimisation principle
underlies the property of self-organisation) and mechanisms (e.g.
non-linear cause and effect).

Principles are important, however; for example, they dictate
taxonomies, hierarchies, rules, causalities, bounds and applicable
contexts, as well as the system behaviours that the theory is about
(i.e. if the theory is about systems).

While non-linearity at the scale of mechanism involves cause
and effect, at the whole system scale non-linearity is a behav-
ioural property, with the whole system yielding non-linear
change that is sometimes entirely unpredictable. So it does not
always hold that neat distinctions between properties and

mechanisms can be made, as outlined earlier.

Indeed, we can establish a tentative hierarchy of concepts that ascribe the complexity of the multiple, constituted of properties, capacities and mechanisms, glued together by principles. But that is as far as we can go. These taxa are themselves capable of morphing from one into another, depending on the context in which they are invoked or arise. They themselves constitute a heterarchy, and that is how it must be for them to constitute a theory that mirrors the reality of the multiple. And this is not vagueness or obscurity; it is dire necessity if an honest theory of the multiple is to be composed. It must not artificially reduce the multiple before encountering it. Rather, it must hold open the space for exploration in and with the multiple, so that new understanding can emerge. That is, the theory should (if possible) facilitate an arrival at a simple understanding or reduction through the contemplation of the complexity of the multiple, but it should not begin with an oversimplification that renders the subject a mere sketch of itself.

The theory should enable us to negotiate a path to simplicity or reduction through a full engagement with the complexity of the multiple. And yet, we cannot ever fully engage with the complexity of the multiple—except perhaps through 'direct experience' of it, in the metaphysical sense—and so we must have good strategies for engaging it which allow us to both understand and acquire a 'feel' for the subject; one that is sensitive to how context and specificity frame the inquiry.

The multiple, or its features and sub-systems, cannot be captured with a single lens. And so it is through the multi-filter—or the multiple as filter—that we must negotiate the complexity of the multiple. It is akin to the homeopathic principle or the central principle of martial arts; we must engage the multiple on its own terms, and not ours, lest we disturb and perturb it away, and end up with an abstraction that bears little or no resemblance to reality. Instead, we must engage with both intellect and feel in order to negotiate complexity. This is unavoidable. As no amount of cutting away the flesh from the bone reveals what makes a person—in all his or her complexity—so too can no amount of reducing the multiple to its elements reveal its splendour. Its splendour emerges from what binds it together, and from how we observe and experience it.

Chapter 16

Emergence and a Changing World

"The world is flat!" is a statement often repeated in this new era. It is a common way of stating that the mutual influence of things upon other things has grown stronger in this era. Yet there is more to the picture, much more than this over-used cliché implies and reveals.

We live in an era where interconnectedness between people, resources, places, activities, organisations, nation states and so forth is greater than it has ever been before in known history. Many systems facilitate this unceasing, growing interconnectedness and its increasing densities. The global economic system is a case in point. It facilitates and exerts influence and control over remote interactions, between people, resources, activities etc. spread far across the globe. The global climate—and its changing behaviour—is another case in point; exerting influence and control on the very same systems that drive changes in it at a global scale. It is simultaneously changing ecosystems, while being changed by them in turn. For example, melting permafrost in the tundras is initiated by rising global temperatures, while escalating that very same effect by releasing potent methane gases into the atmosphere. This 'runaway' effect is an example of positive feedback, amplifying an effect exponentially over time.

This era is also characterised by higher rates of change, and the prevalence of non-linear, discontinuous—or abrupt— changes in global and local systems. Here, again, the evolution of the global economy is no longer a 'steady state' affair. Instead, global economic change is less predictable, more subject to the impact of innovations large and small. Increased connectedness has exacerbated this; for example, changes in the global climate can now wreak devastating impacts within the global economy.

A failed corn harvest in the American mid-west, or wheat harvest in Siberia—due to climate change driven factors such as drought or flooding—can severely impact global food prices and availability, rendering poor households (especially in the developing world) to poverty conditions in a short period of time. This has all made decision-making more difficult for organisations and institutions that have traditionally relied on 'steady state' models of the world to make decisions. One author refers to decision-making in this era as negotiating "organisational white-waters" (Malhotra, 1999).

Technological change, for example, is no longer subject to incremental development trajectories. In the semiconductor, internet and mobile-phone era, the vast potential for innovation has resulted in competitive techno-industries swopping pole positions on an almost annual basis. One year Apple is ahead and Samsung is behind; the next it is reversed; the following year a new competitor has stormed to the lead … it is all difficult to negotiate if one's model of the world presupposes an orderly, linear changing world. A different understanding, strategy and negotiation is required in a world where a youngster's tinkerings in a garage, or on a computer or the internet, can rapidly grow from a niche into a normative everyday feature of our reality.

The spatial changes of our era are also unprecedented in human history. We are quickly advancing into an era where urban populations outnumber rural populations. The cities of the developing world are growing the fastest—i.e. in rates and magnitudes—even though they are not accompanied by rapid industrialisation (except perhaps in the case of China). This has major implications for how societies and cultures will evolve in this era. Moreover, rapid, mostly negative changes in our global ecosystems—which are essential for supporting human and other life—are also unfolding. The Millennium Ecosystem Assessment found that sixty percent of the 24 global life-supporting ecosystem services were being degraded. In the last few decades alone, around fifty per cent of global biodiversity has been lost. Spatio-temporal changes in vegetation—for example, drought and desertification in the Sahel region, which has seen a 200km southward migration of the region—is wreaking havoc in poor countries around the world, where competition over resources such as arable land, grazing land and water often results in bloody conflict.

And as global population pressures grow and intensify— especially in cities—so does the demand for higher living standards. In a world where resources are limited and finite, this requires substantial innovation and change in how society functions, as well as how it perceives itself. What is clear is that consumption patterns are directly linked to the majority of negative change effects that threaten the sustainability of humanity as a whole on planet Earth.

To add to this 'Brownian' milieu of persistent and unpre- dictable change, societies and individuals are also undergoing marked changes. For some, there is a retreat into histories and identities long past, in a desperate attempt to find moorings in this vast sea of change. Others abandon convention and embrace new ways of understanding themselves, each other and their place in the world. Consequently, the way that identity is construed and maintained amidst the varied influences on society and the individual is also undergoing complex changes. It cannot easily be predicted what shape and form society and culture will assume in different places across the planet.

Yet of this era of change, it cannot be said that a break with the past has occurred. For it is through Lazarus that all this change is experienced and negotiated, and the past still manifests strongly within the changing whirlpool of the present. Bounced along eddies and currents, sometimes moving forward, sometimes backward, sometimes inward, sometimes outward, Lazarus is the agent through which all this change is digested and excreted. Lazarus has never been more challenged by uncertainty, and at the same time proffered more choice and opportunity. What we can say about this era, however, is that *emergence*—i.e. that which cannot be predicted—is a central trait of our times. It is all around us and within us; a major source of what constitutes the reality that we reside in and contribute to.

Chapter 17

Polycrisis and Undecidability

And the more interconnected and uncertain the world has become, the more vulnerable we have become to how change manifests and reverberates through its interlinked systems, networks and assemblages. *Combinatorial effects* now have significant potency. No decision can be made in isolation. The decision-making sphere has to be cast wider, and decisions need to trace their potential impacts through a vast array of interconnected systems. And even then, the success of a decision is not guaranteed. This constantly shifting, changing system can easily subvert the most well-thought-through strategy, rendering it defunct by default. Simply put, changes in the multiple can outpace a decision and beat it to the punch, whereupon a decision becomes obsolete or must at the least be adapted or changed to reflect the new unfolding reality.

And even crises combine in this era. Uncertainties in the global economy and climate, for example, can combine to wreak devastating effects at local levels. A resource availability crisis may impact across a range of sectors and activities. In order for whole system sustainability to be achieved, it may be required that the sub-optimisation principle be invoked. It stands to reason that difficult decisions may be necessary to consider, as not all sectors can function at fully optimal levels if sustainability is to be ensured. At times, these decisions may prove veritably undecidable and deeply disputed and contested. It may be that every possible decision proves undesirable to one sector or another, or one actor or stakeholder or another. Deadlock is a fundamentally unavoidable phenomenon when faced with complex challenges that involve and reach across the multiple. That is, the controls through which self-organisation is administered may be disputed, even though it may be commonly understood that a worthy shared objective (such as sustainability, peace and stability, reconciliation) is the ultimate goal.

This vulnerability is, however, at the same time, accompanied by increased potential to influence, and an abundance of choice. That is, Lazarus can embrace antifragility in the midst of it all. This antifragility can emerge from innovation, broader inclusion as well as understanding degeneracy and the various options for adaptation that it enables. Moreover, attaining a shared understanding of these options, so that they can be debated and discussed by all and sundry who are party to the making of decisions, is vitally important. Polycrisis requires broad consensus to address, precisely because it cannot be addressed from one sectoral, disciplinary or other perspective alone. It is forcing the conditions for inclusive governance and politics, as well as development. It is begging for participatory democracy to become a reality, and not just a procedural exercise in governance.

Chapter 18

Heterarchy and the Virtual Realm

Our world, seemingly global, is in reality a planet of thousands of the most varied and never intersecting provinces. A trip around the world is a journey from backwater to backwater, each of which considers itself, in its isolation, a shining star. For most people, the real world ends on the threshold of their house, at the edge of their village, or, at the very most, on the border of their valley. That which is beyond is unreal, unimportant, and even useless, whereas that which we have at our fingertips, in our field of vision, expands until it seems an entire universe, overshadowing all else. Often, the native and the newcomer have difficulty in finding a common language, because each looks at the same place through a different lens. The newcomer has a wide-angle lens, which gives him a distant, diminished view, although one with a long horizon line, while the local always employs a telescopic lens that magnifies the slightest detail.
—*Kapuscinski* (The Shadow of the Sun, *1998*)

It might be said that Kapuscinski's understanding of the world, while reasonable, may be set to change significantly in the 21st century. A fundamental shift is underway in how people everywhere, except the remotest regions, relate to each other and the wider world.

Today, the 'virtual realm' is intervening and disrupting with more impact than ever before. It is more heterarchical than hierarchical in nature, enabling contact with a broad global audience. It also intervenes at the level of the individual and impacts identity significantly. It is a non-trivial development, as the rumblings of the early part of this century have already demonstrated. Its key contribution, at a fundamental level, is its ability to enable *heterogeneity and heterarchy* and to subvert homogeneity and hierarchy in turn, though this dichotomy is a loose one at best. Various degrees of each emerge.

> *The American sociologist James Scott has written wisely of the benefits of what he calls 'local knowledge'. The more variegated and complicated a society, the greater the chance that those at the top will be ignorant of the realities at the bottom. There are limits, he writes, "… in principle of what we are likely to know about a complex functioning order". The benefits of state interventions on the public behalf must always be weighed against this simple truth.*
> —*Tony Judt* (Ill Fares the Land, 2011)

This evolution of a heterarchical virtual realm is–at the same time–a product of the technology, as well as the human need or desire for it. And a key desire, which has found expression through the virtual realm, is the need to connect from the local to the global, as well as from the local to regional and macro-levels of governance.

The need for services—for example, financial services, news, media, advertising, health and agricultural advice and support— that are not readily available in many parts of the developing world, is also increasingly being met through the technologies of the virtual realm. More people now own mobile phones (around 6 billion) than have access to toilets (around 4.5 billion). Herders in rural Somalia can now access regional and global markets through mobile phones and adjust their livestock prices accordingly, as well as find new buyers far beyond their borders. All this in addition to being able to stay in touch with family, friends and clansmen who are spread across the world at the same time. New opportunities, potentials and innovations are already steadily underway in the virtual realm, and are busy at work changing the way we experience the world and our function(s) within it.

And it extends beyond the merely functional aspects of our

existence. The virtual realm enshrines both anonymity and multiple identity, afforded by the interplay between the two. That is, one can only embrace the multiple nature of identity through the ability to be anonymous. Anonymity opens up the space for 'play' with identity, in a manner hitherto unknown. It is not quite the shamanistic multi-identity, which embraces both the physical and spiritual worlds, respectively, as a dichotomy. In this era, anonymity is afforded in the virtual realm for the expression of a variety of identities, all of which may be simultaneously held by the person concerned. The ease with which these identities may be embraced and discarded differs significantly from the shaman's encounter with the spiritual. The virtual world is, after all, just another medium through which real-world interaction is made possible, albeit collapsing space and time tremendously in order to facilitate such interaction.

Heterarchy is intimately related to open-ness or "open systems", a concept that is in turn intimately related to the theory of complex systems, as well as assemblages. In both, complex systems and assemblages are defined as having parts or sub-systems that belong to other systems or assemblages. Hence they become vulnerable to "relations of exteriority" (de Landa, 2006; Deleuze & Guattari, 1987). The complex behaviours that result from this dependence on the behaviours of parts or sub-systems that lie outside the control of the system or assemblage in question, render them vulnerable to unforeseen and uncontrollable change effects[6].

This 'open-ness' is itself a duality, for it enables the conditions for vulnerability of a system or assemblage, but conversely also opens up new potentials for resilience and innovation. It is a matter of orientation, perspective, and adaptive and creative capacity that determines which prevails.

Chapter 19

Transdisciplinarity and the Future

The moment we turn our mind to the future, we are no longer concerned with "objects" but with projects, and it is not decisive whether they are formed spontaneously or as anticipated reactions to future circumstances. And just as the past always presents itself to the mind in the guise of certainty, the future's main characteristic is its basic uncertainty, no matter how high a degree of probability prediction may attain. In other words, we are dealing with matters that never were, that are not yet, and that may well never be. Our Last Will and Testament, providing for the only future of which we can be reasonably certain, namely our own death, shows that the Will's need to will is no less strong than the Reason's need to think: in both instances the mind transcends its own natural limitations, either by asking unanswerable questions or by projecting itself into a future which, for the willing subject, will never be.
—Hannah Arendt (The Life of the Mind, *1971)*

Our ways of thinking about and addressing complex, integrated challenges—i.e. the "grand problematiques" of our era (Max-Neef, 2005)—must, by necessity, reflect the complexity of the problem or challenge being addressed. Reflecting and addressing this complexity, in turn, requires that multiple perspectives—in particular, disciplinary perspectives—be brought to bear on the integrated challenge at hand. That is, a *transdisciplinary* approach is required.

Transdisciplinarity is more than integration of many disciplines under one of them—i.e. interdisciplinarity—it is the fully heterarchical integration of disciplines. This integration can be thought of hierarchically—as Max Neef has envisaged it—where a hierarchy of disciplines can be conceived of, and a transdisicplinary action is that which traverses the different levels of the hierarchy. It can also be conceived of non-hierarchically, where instead of induction and deduction being deployed in service of understanding a particular system abduction is used instead. As previously outlined, abduction occurs where a concept or metaphor from one field or discipline is applied in another discipline, and reveals new insights and interdependencies that were hitherto undiscovered. It is a little-known fact that the majority of scientific and other great theoretical leaps are in fact products of the process of abduction, rather than the more commonly understood scientific processes of induction and deduction (Peirce, 1934; in Hodgson, 1993).

In an alternative conception, transdisciplinarity is defined as systems knowledge, target knowledge and transformative knowledge, which distinguishes it from theories that are concerned solely with understanding (and not influencing) the subject. In this conception of transdisciplinarity the transdisciplinarian is concerned with how their understanding of a particular system or problem enables them to envision and take actions to influence the subject of inquiry.

In summary, there are three ways of conceiving of transdisciplinarity: (1) a transdisciplinary action that cuts across disciplinary hierarchies as laid out by Max-Neef (2005); (2) employing abduction of concept or metaphor in order to yield knowledge for new insights and innovations; and (3) formulating knowledge for intervention and action.

These three ways of conceiving of transdisciplinarity are not mutually exclusive, and it is easy to envisage that some overlap between them exists. All are relevant where envisaging strategies for change is concerned, but likely address different aspects of formulating pathways into the unknown and venturing into it.

The success of strategies for coping with the complexities of this era will likely increasingly depend on our ability to integrate across disciplinary silos and hierarchies and realise integrated programmes of action that seek to address the grand problematiques of this era. It is clear that 'more of the same' reductionist and/or mono-disciplinary approaches are, and have always been, unable to cope with the complexities of the multiple. They have only proved useful when change was relatively predictable, linear and reliable (i.e. in rare moments or periods in history). They fall flat when faced with variability, non-linearity and uncertainty.

And so it is with transdisciplinarity that it is heavily concerned with moving from a theoretical focus on "objects" to "projects"—as Hannah Arendt (1971) articulates it—and it must also hence be subject to the exercise of the "will". That is, reason is not enough. Simply put, the world does not change just because it makes sense to. Change is a product of reason and will, but it is also fundamentally emergent as it emerges from real-world systems that do not necessarily always act reasonably.

And it is to the will that we must turn to in order to realise purposiveness, to go beyond the calculable, into the realms of the future. Yet the will is useless without adequate awareness, and Lazarus must negotiate both in order to move into and shape the future. Lazarus as agent of change must mobilise to act in concert with others in order to bring about broader changes within the multiple. As transdisciplinarian, Lazarus must be able to converse in the different languages that establish artificial divisions within society. He must actively participate in co-constructing a shared understanding between opposing and varying perspectives. To realise this, Lazarus must abandon one-dimensional analyses and closed, self-referential languages (i.e. of ideologies, disciplines, sectors, institutions, etc.), and seek out a new way of seeing; one that is independent of mere metrics, and strict hierarchies and taxonomies. And this new way of seeing requires that we abandon mechanism as the starting point, and look to the properties of the multiple as a medium for engendering shared understanding and debating our future.

Chapter 20

Final Note

In every revolution, a movement grapples with a structure. The movement attacks the structure, trying to destroy it, while the structure defends itself and tries to extinguish the movement. The two forces, equally powerful, have different properties. The properties of a movement are spontaneity, impulsiveness, dynamic expansiveness—and a short life. The properties of a structure are inertia, resilience, and an amazing, almost instinctive ability to survive. A structure is rather easy to create, and incomparably more difficult to destroy. It can long outlast all the reasons that justified its establishment. Many weak or even fictitious states have been called into being. But states, after all, are structures and none of them will be crossed off the map. There exists a sort of world of structures, all holding one another up. Threaten one and the others, its kindred, rush to its assistance. The elasticity that helps it to survive is another trait of structure. Backed into a corner, under pressure, it can suck in its belly, contract, and wait for the moment when it can start expanding again. Interestingly, such renewed expansion always takes place exactly where there had been a contraction. Structures tend toward a return to the status quo, which they regard as the best of states, the ideal. This trait belies the inertia of structure. The structure is capable of reacting only according to the first program fed into it. Enter a new program— nothing happens, it doesn't react. It will wait for the previous program. A structure can also act like a roly-poly toy: Just when it seems to have been knocked over, it pops back up. A movement unaware of this property of the structure will wrestle with it for a long time, then grow weak, and in the end suffer defeat.
—*Kapuscinksi* (Shah of Shahs, 1982)

The filters proposed here are not exhaustive, but they provide a way of moving beyond analyses that stem from a mere understanding of mechanism, or from a closed disciplinary framework (such as a cost-benefit analysis in economics). Rather, they open up space for interrogating the complexities of the challenges we face, and envisaging new possibilities for the future.

This is not a theory of revolutions, but of 'evorevolution'. It is a theory of how to negotiate the past, present and the future using a wholly new framework of thought. Its aim is not to dismantle whole systems and replace them with new systems, which for all their utopian ideals quickly diverge from their initial conditions as they encounter the reality of the multiple. Rather its aim is to invoke a more subtle, yet profound, method of engaging the socio-political and cultural fabric in which we live. Its aim is to enable an evolutionary perspective on change, to allow for new possibilities and potentials to seed, grow and thrive—if it has purchase within the multiple—and to deliberately avoid adopting a positivist stance on societal change. To repeat, we cannot engineer social change, except through authoritarianism and totalitarianism, and even then, this engineering cannot be sustained in the world we live in today.

Instead of brute force, subtlety and novelty are called upon as vessels and catalysts for change. Moreover, the approach tendered in this book does not depend on a benevolent 'intelligentsia' to set out the pathways and trajectories of society, on behalf of the masses. Rather, it calls upon Lazarus to awaken, to move in awareness, to act upon the will, in order to connect with others and mobilise for change. It seeks to empower the disempowered, through providing a means for their inclusion in negotiating the future and participating in realising it. It is not a moral theory, and neither is it solely empirical. It is simply a theory that attempts to engage with the multiple on its own terms instead of the terms we seek to manage our narrow interests in the world.

Its usefulness will undoubtedly depend on who is wielding it, and how well they are able to make use of the conceptual framework tendered here. It is not a theory that simply seeks to provide calculable answers to the complex and difficult questions we face in negotiating the 21st century. It is a theory that allows deeper exploration into complex phenomena, and opens up potentials for influencing the trajectory of the multiple, for participating in that evolution, rather than controlling it. It is a strategic framework for engaging complexity, and not a formalism for deriving neat abstractions of complexity.

It does not prescribe political, economic and/or other systems that constitute the fabric of the multiple. Rather, it acknowledges that in this moment in history there is a need to allow for new systems to be conceived of, debated, discussed and tested. Moreover, that this should be a continual practice is central to this approach; we should never be fooled into thinking that any system is cast in stone, immovable. All permanent structures face dissolution as soon as their permanence is announced; as it is precisely at this moment that the structure begins to fossilise, to resemble an artefact of itself. It ceases to breathe and evolve with the multiple, and becomes a constraint upon it instead. Held back by this outdated structure within it, the multiple exerts its own pressures upon it, and if it cannot adapt, evolve, it eventually breaks under the pressures and is left to ruin, for future generations to gaze at and wonder at the impermanence of all constructions.

And so, this theory is not about establishing utopian systems that will run ceaselessly into the future without hitch. It is about acknowledging the fraught but liberating nature of change: that while it surprises and catches us unawares it also opens up potentials for new futures. It is this duality that we must reside in and act from if we are to negotiate the complexity of the multiple. And being neither black nor white, it is an uncomfortable position. But there is great benefit in this discomfort if it is understood for what it is: the progenitor of all reality and change. Its power is that it does not prescribe any single over-arching theory, but enables an approach founded in a set of properties and principles that are not absolute but recognise paradoxes, contradictions and undecidables as part of everyday reality. Lazarus must awaken to this perspective, in order to develop the sensitivity that is required for engaging complexity. And so we must look to the *artist* to understand how this sensitivity develops, so that Lazarus can break free of prescriptions and engage the multiple with open eyes.

> *Expressive action begins with sensing a rupture in existence. The desire to eliminate this gap and become fused with existence itself becomes the will to create art.*
> —*Lee Ufan* (Marking Infinity, 2011, Guggenheim)

It becomes equally important, if not more important, to develop a sensitivity than to develop a language. This sensitivity is more about going within than going without. It is to go deeper into what is felt rather than what is thought. A language that is born of this sensitivity does more than just communicate; it resonates. It can be felt as well as understood. It can be both specific and universal at the same time. Akin to music, it is not a linear set of logics that make the pretence of objectivity; rather it is a subtle movement that acknowledges the irreducibility of the multiple, and attempts to meet it on its own terms, to participate in and with it, and to evolve with it.

Reductionism and abstraction work in different directions, but attempt the same feat. Yet there are no essences and totalities within the multiple but those that we ascribe to it. We see the boundaries, but it is entirely likely that the multiple has no 'knowledge' of this, that its structures are not evident to it, except as limiting conditions. In the realm of the temporal, all is imper-manent, and that is the 'essence' of complexity; that what might be thought to be cast in stone inevitably changes. We cannot control these changes but temporarily, and so in order to look to the long-term and act upon it we must act with the multiple and acknowledge its vagaries and complexities. And even if it itself is purely a construct of our own making, a reflection of what is within, there is great value in acknowledging it nonetheless. In doing so we cease to become mere subjects of change, powerless within it. Instead, we become what we are in truth: participants in the multiple that shape it as it shapes us. From this reality there is no escape, and hence, this truth constitutes the starting point for the theory of the multiple that is proposed in this book.

Afterword:
Reflections on Descartes' *Meditations*

In Descartes' first meditation—aptly named, "things which can be called into doubt"—he embarks on what he hopes will prove to be a process of liberation (Descartes, 1641). He is acutely aware that there are beliefs that he has held true, but which may not be; and hence require deep scrutiny. There is a sense of a final confrontation between Descartes and his own beliefs; one that will fundamentally change his knowledge of his place in the world, and the world with it. And so we must chart and follow Descartes' journey carefully, lest we lose the value in it.

In the search for 'truth', Descartes attempts to refute all his previous beliefs, if there is even a hint of anything that can be doubted about them. Every belief, thus dismissed, will collapse the structure of artifice and only that which is certain and true will remain, or so he hopes. He states his intentions clearly:

> *But since reason already convinces us that we should withhold assent just as carefully from whatever is not completely certain and indubitable as from what is clearly false, if I find some reason for doubt in each of my beliefs, that will be enough to reject all of them. However, they need not all be reviewed individually, for that would be an infinite task; as soon as foundations are undermined everything built on them collapses of its own accord, and therefore I will challenge directly all the first principles on which everything I formerly believed rests.*

And thus, Descartes attempts to reveal absolute truth by doubting everything that he has previously believed in. But absolute truth, in itself, can exist only—as Descartes understands it—because God exists; the perfect designer who holds absolute truth in His hands, the truth from which the laws of the universe and all creation arises. His campaign of doubt, fuelled by reason, can only be undertaken in the safe knowledge that there is an ultimate destination for such a quest, and that destination, being ultimate truth, requires some ... *thing* that understands it—there can be no knowledge without an observer—and only God can possibly hold and comprehend such knowledge, being infinite Himself or Herself. Descartes invokes absolute truth and God, in order to make sense of his quest. The methodology he adopts in undertaking his exploration of metaphysics is profoundly affected by his ability to hold this duality in play during his meditations. Doubting everything *necessitates* the invocation of "God" and absolute "truth", for if they do not exist then his search is in vain. He assumes that the object—truth—exists. And so with God. Indeed, it is his starting point:

> *First one needs to know that all substances—that is, things which, in order to exist, have to be created by God—are without exception incorruptible by their nature, and they can never cease to exist unless they are reduced to nothingness by the same God if he stops maintaining their existence.*

The irresolute fact of his existence is, by his own admission, reflected in his belief in God. Just as man has dominion over nature—biblically speaking—so too does God have domain over the universe and everything within it. In Descartes' conception only God can render nothingness unto beings, where nothingness is no longer an attribute of an entity that exists, but rather, the entity itself is "reduced to nothingness". Hence God is rendered to the realm of control over absolutes, while *we*—mere mortals—are consigned to the realm of the relative, adrift within the multiple, the complex. Descartes' quest is that of metaphysics, and both reason and faith coexist in his meditational endeavours.

He accepts the possibility, however, that his meditational

quest may well ultimately end in vain, stating: "I will follow this strategy until I discover something that is certain, or at least, until I discover that it is certain only that nothing is certain". He accepts the possibility of failure; his meditation is an experiment. He goes the opposite direction of Serres in his quest, however, denying senses and emotion wholesale:

> *Thus I will assume that everything I see is false. I believe that, among the things that a deceptive memory represents, nothing ever existed; I have no senses at all; body, shape, extension, emotion, and place are unreal. Perhaps that is all there is, that there is nothing certain.*

He reacts to uncertainty with a scalpel. He attempts to dispel all that is uncertain, even in small part, so that the certain can emerge in his meditation. He does the opposite of what meditation is about; that is, to pull yourself into the present and grow in awareness of how thoughts and emotions manifest. Descartes' meditation, by contrast, is not designed to still the mind and enter the present fully, with all senses intact. Descartes' meditation is designed to still the senses instead of the mind. The mind continues unabated. And as it is with the yoga of discrimination, where enlightenment is pursued through knowledge, an existential breakdown threatens Descartes[7]. Perhaps he is significantly moved by it, and that is why he writes it out; in order to make sense of it—to obtain perspective on it—and to give the process he is undergoing a sense of purpose.

The yoga of discrimination proceeds by stilling the mind, as well as the senses, and pursuing knowledge of what is changing and unchanging. It is very similar to the process that Descartes has undertaken. Yet whereas the yoga of discrimination begins by acknowledging duality as resident in reality, Descartes begins from the opposite perspective, seeking clear dualism in everything but his invocation of both reason (i.e. the rational) and God (i.e. the irrational) as premise for his meditation. Otherwise, he neatly divides mind and body, man and nature, and so forth. Whereas a core goal of the yoga of discrimination is to determine that the 'I' is itself illusory, Descartes asserts that the 'I' is the one thing he can be certain of in his (second) meditation.

He remains unaware of how uncertainty is generated, and how it manifests, however. He does not consider that it may be resident in reality, indivisible from it, because Descartes works from the assumption that perfection exists. In his framing, God and absolute truth are that in which there can be no doubt, whereas everything else can be doubted. *The truth would emerge as that without doubt.* And if any belief can be doubted, even in part, it may be thus discarded wholesale, for the sake of the meditation. Already, Descartes is adopting the path of the zealot, for he is lapsing into an extremist interpretation of reason itself. In short, what he attempts is in fact unreasonable, but he knows that the consequences of this are minimal, and so can be pursued with vigour and purpose:

> *Therefore, I think I shall not act badly if, having turned my will around in the opposite direction, I deceive myself and pretend for a while that these beliefs are completely false and imaginary until at length, as if I were balanced by an equal weight of prejudices on both sides, no bad habit would any longer turn my judgement from the correct perceptions of things. For I know that no danger or error will result from this in the meantime, and that I cannot exaggerate my cautiousness since I am concerned here not with doing things but merely with knowing them.*

In the language of physics, Descartes is invoking the freedom of the *Gedanken* — or "thought experiment" — so that the meditation allows for deep exploration of the "knowing" of things, and remains unconcerned with "doing".

By doubting everything, he is treating beliefs as though they are themselves 'parts'—presumably of absolute truth. They are seamless, in this sense, with truth, as truth collapses when they—the "foundations"—are "undermined", invalidated:

> *Everything that I accepted as being most true up to now I acquired from the senses or through the senses. However, I have occasionally found that they deceive me, and it is prudent never to trust those who have deceived us, even if only once.*

Yet the senses are perceptual and so the beliefs they help manifest are not parts or components of truth; they are merely partial reflections of truth, where truth is captured momentarily and packaged, ordered, made palatable. And thus, Descartes' beliefs do in fact become mere parts. They can have either a material or an expressive role, but not both.

That is vastly different to the proposition made in assemblage theory (i.e. that a component can play an expressive and/or material role along a continuum), or that made in complexity theory (i.e. that all 'parts' of a system can be treated as agents, and hence also enjoy the ability to play both expressive and material roles; both along a continuum and as duality). The beliefs that Descartes refers to are treated as discrete in nature; they are building blocks of a perfect 'machine' of truth. They are the parts as well as the interdependencies—the structure—that enables the machine of truth to manifest. And in so deciding, Descartes regards the truth in the same way as he regards the body: as merely composed of parts.

To Serres, however, the body is not merely a machine, made up of parts; it is an instrument—i.e. "the body instrument"—through which the multiple is both perceived and negotiated, and is comingled with the senses, emotion, reason, belief, faith, judgement and decision-making.

But to Descartes, the quest for absolute truth—which is in itself a religious pursuit—lies at the heart of his endeavour.

Descartes attempts to approach truth through beliefs; frankly, in finding a perfect belief or set of beliefs. Beliefs, however, are only ever asymptotic to the 'truth', as it were; in theories of complexity and assemblage, as well as a range of others. And in this casting, truth becomes infinite, only reachable through the mystical experience of it. It is not that truth does not exist, only that it cannot be approached through the discrete; it can only be reached through continuity. And this continuity can be provided only through *direct experience in the present*; it cannot be provided by the discrete, as that would be discontinuous with the present. The present knows no absolutes, only uniqueness, and so it and only it harbours direct experience.

The perpetual race of Achilles and the tortoise has a role to play in these considerations as well. And the infinitesimal, whether large or small in distinction, is about the asymptotic; something that can be approached but never joined with. We can approach nothingness (i.e. zero or '0') with the same success that we can approach infinity (i.e. '1/0'). That is, the infinitesimal and asymptotic are absolutes.

The truth is made discrete when broken down into beliefs. It is reduced. It is no longer absolute truth of any kind. It has been captured, put asunder to a hierarchy—put in its place—so to speak. This is what Descartes does. He discards partial beliefs in the quest for truth. He does not consider that all beliefs can only ever remain partial, because truth, once captured, is *reduced* to belief.

The elevated 'truth' which he seeks cannot be attained through the undertaking of discretely discarding beliefs. The truth of which Descartes writes can only be approached through continuity, as it is infinite, containing everything absolutely, purely, perfectly; like the God he imagines.

Herein there is no room for duality. There is no room for an included middle. Everything is binary; true or false, evil or virtuous, and so forth. And hence there is no room for undecidability either; everything must be decidable, as in order for him to pursue the meditation, every belief must be decided upon.

This extreme experiment renders even that which may be truthful in many contexts, but questionable or doubtful in only a few contexts, as *wholly doubtful*. He takes a mix and reduces it to an essence, rightly or wrongly, for the purpose of his meditation. And immediately, the complex—the multiple—is lost, and he is set adrift.

Take *uncertainty*, for example; a property of the multiple. Uncertainty is resident within the multiple, a condition of it, precisely because the multiple is itself a duality. That is, the multiple is both perceptual and material at the same time. Moreover, even when considered in its material sense, it is both ontological and phenomenological. This renders the material as duality itself, as well as being a participant in the duality of the multiple at the same time. And so, emotion, sense and reason are inextricable. They entangle and manifest contradictorily at times, coherently at others.

Moreover, *undecidability* cannot be that easily reduced. A professional fighter, for example, lives to fight yet also fights to live. Both must be embraced and integrated in the fighters' quest. The former invokes freedom, yet the latter the need for control. Is he therefore a walking contradiction? Yes, but that is the condition of being human: that we are multiple. And for the fighter, both freedom and control are required to co-exist, whether uneasily or not.

Undecidability can sometimes manifest as *duality* (when something is two or more things at the same time, whether these 'things' constitute material components, judgements, concepts, language, etc.)[8], and sometimes it can manifest as *entanglement* (when opposites manifest a dependence on each other for both their mutual existence and their mutual destruction[9] e.g. such as the aforementioned combination of 'control' and 'freedom' that a fighter requires in order to function effectively).

Presumably, it can also manifest as a result of both processes unfolding at the same time, i.e. an undecidable may exhibit duality. In this case, as is the case with particle physics, it would constitute 'duality all the way down', i.e. duality would be replicated at finer and finer scales of resolution and measurement.

Yet this is not restricted to the world of particles. By way of metaphor, it has relevance in social systems as well. While Descartes attempts to separate mind from body, and reason from emotion and sense, he does so hesitantly. He is unsure whether it is at all possible. He is, for all intents and purposes, undecided, but feels compelled to decide one way for the sake of his meditation. As he reveals:

> *However, I have already denied that I have any senses or any body. I still cannot make any progress, for what follows from that? Am I so tied to a body and senses that I am incapable of existing without them?*

And on distinguishing between the world of dreams and the waking world he is similarly disposed, in no small part due to his insomnia. He is concerned with the question of reality, and for him, it cannot be anything about which there can be any doubt. As he puts it:

> *When I think about this more carefully, I see so clearly that I can never distinguish, by reliable signs, being awake from being asleep, that I am confused and this feeling of confusion almost confirms me in believing that I am asleep.*

For he struggles to reduce the multiple, precisely because he is both multiple within and without. He attempts to filter multiplicity in an enduring but misguided quest to reduce it to its truths. He does not make Serres' observation: that the senses are knotted, entangled, indistinct, and therein lies their profound, even irrational power.

And according to more recent studies (Damasio, 1994), it also seems that reason, judgement and emotion cannot be separated neatly either. We rely on emotions to reason and make sound judgements. We live with a different understanding of the world than prevailed in Descartes' time. Yet, if it is true that in Lazarus, reason, judgement, emotion and sense are all entangled, it is also true that it was the same in René Descartes.

He grasps at pure, mathematical, abstract reason to convince himself of his cause, so that he can escape this entanglement:

> For whether I am awake or asleep, two and three added together make five and a quadrilateral figure has no more than four sides. It seems impossible that one could ever suspect that such clear truths are false.

But there is no escaping the inevitable now. Descartes must invoke the only belief that can save his quest, for adrift in the multiple he finds no solace, has no filters, no understanding with which to engage it. He has taken the path of extremes, the first of which being that of pure reason, and so it is only in the other extreme—faith—that he can harness the irrational into something meaningful:

> However, there is an ancient belief somehow fixed in my mind that God can do everything and that I was created by him with the kind of existence I enjoy. But how do I know that, although he created absolutely no earth, no sky, no extended things, no shape, no magnitude, no place, he still arranged that all these things would appear to exist, as they currently do?

He looks to faith—i.e. masquerading as an "ancient belief"—to answer his call. He looks to God, and in doing so he is bound to invoke the "devil" as well, a devil—"some evil mind"—that deceives him, works through the senses to confuse him, rendering him unto the fate of Lazarus, caught in the multiple:

> *Therefore I will suppose that, not God who is the source of truth, but some evil mind, who is all powerful and cunning, has devoted all their energies to deceiving me. I will imagine that the sky, air, earth, colours, shapes, sounds and everything external to me are nothing more than the creatures of dreams by means of which an evil spirit entraps my credulity.*

Hence Descartes engages metaphysics with reason at first, but has to retreat to faith—*the irrational* in Kierkegaard's conception—in order to save his quest. He has pre-diagnosed the evil as emerging from a cunning spirit that works through the senses to obscure his vision of the truth, of God itself. He makes a mighty effort, but there is an admission that the effort is too great to bear in daily reality:

> *I shall imagine myself as if I had no hands, no eyes, no flesh, no blood, no senses at all, but as if my beliefs in all these things were false. I will remain resolutely steady in this meditation and, in that way, if I cannot discover anything true, I will certainly do what is possible for me, namely, I will take great care not to assent to what is false, nor can that deceiver—no matter how powerful or cunning they may be—impose anything on me. But this is a tiring project and a kind of laziness brings me back to what is more habitual in my life. I am like a prisoner who happens to enjoy an imaginary freedom in his dreams and who subsequently begins to suspect that he is asleep and, afraid of being awakened, conspires silently with his agreeable illusions. Likewise, I spontaneously lapse into my earlier beliefs and am afraid of being awakened by them, in case my peaceful sleep is followed by a laborious awakening and I live in the future, not in the light, but amid the inextricable darkness of the problems just discussed.*

And so Descartes, without awareness, remains trapped in the same predicament as Lazarus, caught between awakening and the illusory. And as he has divided God and the devil, mind and body, man and nature, he himself becomes divided. Unable to raise a clear signal, or distinguish one, he is trapped within a hell of his own making. The "cunning deceiver" is no other than the multiple within; *as it is within, so it is without!* The noise of the multiple converges within and upon him and he is helpless.

Having begun his meditation by excluding the possibility of the included middle (i.e. duality), Descartes has only *dualism* with which to negotiate the complexity of the multiple. Yet, *duality*—the opposite of dualism—is also self-replicating, another fundamental property of the multiple itself, as well as all that resides within it. He cannot negotiate the multiple effectively without it. Discarding duality, as a key filter, renders Descartes consigned to the same weary path as Lazarus.

Without tools to filter the multiple, to understand it and act upon it, Descartes becomes Lazarus, with only his propensity for reduction to anchor him in the vast waters of the multiple. He is brilliant, yet blind; and his contribution is rather that he plunged so readily into the waters of the multiple that he drew attention to it, so that one day it may be revealed in all its complex glory, and pondered upon, so that we may become liberated within it, rather than from it.

Endnotes

1. Especially reductionist, positivist and humanist philo-sophical beliefs, which persist and prevail in decision-making in many sectors of society, despite their obvious shortcomings in addressing societal challenges adequately.

2. An 'undecidable' falls between polar definitions that seem irreconcilable. Many real physical phenomena, such as light—which is both a wave and a particle at the same time—exhibit *duality*, and not *dualism*, in how they manifest in reality. This is described in more detail in Chapter 12, "The Multiple as Filter".

3. A heterarchy is an adaptive hierarchy, and can be envisioned as a flattened fishnet wherein different controls, functions, processes, agents etc. rise to authority, depending on the particular context of that moment.

4. "That is, a strategy should open up spaces of opportunity, like a boxer; feinting and adjusting so as to open up oppor-tunities in the future—not aiming direct, singular blows." Refer to Chapter 3, "The Wakeful Dead" in Part I.

5. This is also true of the act of observing, and self-organisation is contingent on the filters or 'ways of looking' at a system (Gershenson and Heylighen, 2003).

6. In this conception, complex systems or assemblages are coherent/consistent, but never complete, satisfying—in a sense—the mathematical "uncertainty principle"; that is, Godel's theorem, which states the same about mathematical sets.

7. Jnana Yoga is the yoga of discrimination, which involves engaging in discriminating what is never changing from what is illusion. The proponents constantly assesses 'not this' and 'not that' in order to discriminate the unchanging from the illusory. The 'self' also becomes regarded as

illusory, but this often necessitates an existential crisis, so that a breakdown—or breakthrough—can occur, leading the proponent to a new, enlightened understanding of themself that is liberated from the ego or self.

8. That is, duality is when a *single* thing (e.g. object, entity or decision) has two attributes, both of which constitute metaphysical opposites (e.g. light is a 'waveparticle' or a decision can be good or bad, etc.).

9. In the case of entanglement, *two things* manifest in relation to each other, and are coupled metaphysical opposites that define or cannot exist without each other. Many such couplings can manifest in reality (for example, in a system or assemblage).

References

Arendt, H. (1958) *The Human Condition*, Chicago, University of Chicago Press.

Arendt, H. (1961) *Between Past and Future*, New York, Penguin Books.

Arendt, H. (1971) *The Life of the Mind*, USA, Harcourt.

Arendt, H. (1978) *The Jew as Pariah*, New York, Grove Press, Inc.

Baas, N. A., and Emmeche, C. (1997) 'On emergence and explanation', *Intellectica*, vol. 25, pp. 67–83. Available at http://www.nbi.dk/~emmeche/coPubl/97d.NABCE/Expl Emer.html (Accessed on 29 April 2015).

Banisch, M. (2002) 'Derrida, Schmitt and the essence of the political', *Political Theory Stream Jubilee Conference (refereed paper)*. Australasian Political Studies Association, Australian National University, Canberra, Australia, October 2002 [Online]. Available at https://www.academia.edu/2031268/Derrida_Schmitt_and _the_essence_of_the_political (Accessed 28 April 2015).

Bankoff, G. (2001) 'Rendering the world unsafe: "Vulnerability" as Western discourse', *Disasters*, vol. 25, no. 1, pp. 19–35, March.

Biko, S.B. (1987) *I Write What I Like*, Johannesburg, Heinemann.

Borges, J.L. (1944) *The Perpetual Race of Achilles and the Tortoise* (trans. E. Allen, S. Jill Levin, E. Weinberger), United Kingdom and New York, Penguin Books (this edition 2010).

Cilliers, P. (1998) *Complexity and Postmodernism*, United Kingdom and New York, Routledge/Taylor and Francis (this edition 2002).

Cilliers, P. (2001) 'Boundaries, hierarchies and networks in complex systems', *International Journal of Innovation Management*, pp. 179–186.

Damasio, A. (1994) *Descartes' Error: Emotion, Reason and the*

Human Brain, London, Vintage (this edition 2006).

De Landa, M. (2006) *A New Philosophy of Society: Assemblage and Social Complexity*, New York, Bloomsbury Academic.

Deleuze, G.; Guattari, F. (1987) *A Thousand Plateaus. Capitalism and Schizophrenia*, Minneapolis and London, University of Minnesota Press.

Derrida, J. (2001) 'On forgiveness', in *Cosmopolitanism and Forgiveness*, Thinking in Action Series, Routledge, Available online at http://www.columbia.edu/itc/ce/s6403/jacques_derrida.pdf (Accessed 28 April 2015).

Derrida, J. (1992) 'Force of law: "The mystical foundation" of authority', in Cornell, D. Rosenfeld, M., & Carlson D. (eds.), *Deconstruction and the Possibility of Justice*, New York, Routledge.

Descartes, R. (1641) *Meditations*, London and New York, Penguin Books (this edition 2010).

Gandhi, M.K. (1922) *Ethical Religion*, Ahmedabad, Navajivan Publishing House (this edition 1968).

Gershenson, C., & Heylighen, F. (2003) 'When can we call a system self-organizing?', in Banzhaf, W. Christaller, T. Dittrich, P., Kim, J.T., & Ziegler, J. (eds.) *Advances in Artificial Life, 7th European conference. Dortmund*, Springer, pp. 606–614.

Gosling, D.L. (2007) *Science and the Indian Tradition: When Einstein met Tagore* (Series: *India in the Modern World*, Book. 3), London, Routledge.

Heylighen, F., Cilliers, P., & Gershenson, C. (2007) 'Complexity and philosophy', in Blogg, J., & Geyer, R (eds), *Complexity, Science and Society*. Radcliffe Publishing: Oxford, United Kingdom.

Heylighen, F., Joslyn, C., & Turchin, V. (2001) *Principia cybernetic web* [online], Available at http://pespmc1.vub.ac.be/asc/HETERARCHY.html (Accessed 28 April 2015).

Hodgson, G. M. (1993) *Economics and Evolution: Bringing Life Back Into Economics*, USA, University Of Michigan Press.

Hustveldt (2008)) *The Sorrows of an American*, USA, Henry Holt and Company, pp. (130–131). Cited in Van Marle, K. (ed) 'Refusal, Transition and Apartheid Law', Stellenbosch, Sun Press (this edition 2009).

Islam, G., Zyphur, M., J., Beal, D., J. (2006) *Can a whole be greater than the sum of its parts? A critical appraisal of emergence* [Online], IBMEC Working Paper–WPE-11-2006, Sao Paulo. Available at
http://citeseerx.ist.psu.edu/viewdoc/download?doi=10.1.1.318.5653&rep=rep1&type=pdf (Accessed 28 April 2015).

Judt, T. (2011) *Ill Fares the Land*, London and New York, Penguin Books.

Kapuscinski, R. (1982) *Shah of Shahs*, London and New York, Penguin Books (this edition 2006).

Kapuscinksi, R (1998) *The Shadow of the Sun: My African Life*, London and New York, Penguin Books (this edition 2002).

Lawrence, T.E. (1936) *Seven Pillars of Wisdom*, London, Black House Publishing Ltd.

Levy, D., & Sznaider, N. (2005) 'Holocaust', In Ritzer, G. (ed.), *Encyclopaedia of Social Theory* [Online], (pp. 379–382), Thousand Oaks, CA, SAGE Publications. Available at
http://dx.doi.org/10.4135/9781412952552.n144 (Accessed 28 April 2015).

Malhotra, Y. (1999) 'Knowledge management for organizational white-waters: An ecological framework', *Knowledge Management*, vol. 2, pp. 18–21. Available at
http://brint.org/WhiteWaters.pdf (Accessed on 29 April 2015).

Max-Neef, M.A. (2005) 'Foundations of transdisciplinarity', *Ecological Economics*, vol. 53, pp. 5–16.

Peirce, C. S. (1934) *Collected Papers of Charles Sanders Peirce Volume 5*, Hartshorne, C. and Weiss, P, (eds.), Cambridge Massachusetts, Harvard University Press, p. 156.

Peter, C. (2011a) 'Obstacles to a third way: Ideology before analysis', *Thought Factory*, 2 November 2011 [Blog]. Available at http://thoughtfactory-cam.blogspot.com/2011/11/obstacles-to-third-way-ideology-before.html (Accessed 28 April 2015).

Peter, C. (2011b) 'How left meets right and collides into annihilation', *Thought Factory*, 22 October [Blog]. Available at http://thoughtfactory-cam.blogspot.com/2011/10/how-left-meets-right-and-collides-into.html (Accessed 28 April 2015).

Peter, C. (2011c) 'The Occupy protests: Being leaderless helps reshape the territory', *Thought Factory*, 5 November [Blog]. Available at http://thoughtfactory-cam.blogspot.com/2011/11/occupy-protests-being-leaderless-helps.html (Accessed 28 April 2015).

Piketty, T. (2014) *Capital in the 21st Century*, London, Bellknapp Press of Harvard University Press.

Resnikoff, N. (2014) 'Hunger and homelessness rise in several US cities', *Al Jazeera Online*, 11 December [Online]. Available at http://america.aljazeera.com/articles/2014/12/11/hunger-and-homelessnessriseinseveraluscities.html (Accessed 28 April 2014).

Richardson, K.A. (2005) 'Systems theory and complexity', in Richardson, K.A., Snowden, D., Allen, P.M. and Goldstein, J.A. (eds), *Emergence: Complexity and Organization 2005 Annual*, part 3, vol. 7, pp. 104–144.

Sartre, J. (1965) *Nausea*, United Kingdom and New York, Penguin Books.

Serres, M. (1982) *Genesis* (trans. G. James and J. Nielson), USA, University of Michigan Press (this edition 1995). Originally published in French by Editions Grasset et Fasquelle.

Serres, M. (1998). *Les Cinq Sens*, Paris, Hachette Littératures. Translations of quoted text by Connor, S. (1999) 'Michel Serres's five senses' [Online]. Available at http://stevenconnor.com/5senses.html (Accessed 28 April 2014).

Skyttner, L. (2001) *General Systems Theory: Ideas and Applications;* River Edge, New Jersey (USA), World Scientific.

Taleb, N.N. (2012) *Antifragile,* London and New York, Allen Lane/Penguin Books.

Thaler, R.H, and Sunstein, C.R (2008), *Nudge,* London and New York, Penguin Books (this edition 2009).

Žižek, S. (2009) *First as Tragedy, Then as Farce,* London and New York, Verso.

Žižek, S. (2011) 'Slavoj Žižek: Capitalism with Asian values, *Al Jazeera Online* [Podcast]. 13 November. Available at http://www.aljazeera.com/programmes/talktojazeera/2011/10/2011102813360731764.html (Accessed 28 April 2015).

Contemporary culture has eliminated both the concept of the public and the figure of the intellectual. Former public spaces – both physical and cultural – are now either derelict or colonized by advertising. A cretinous anti-intellectualism presides, cheerled by expensively educated hacks in the pay of multinational corporations who reassure their bored readers that there is no need to rouse themselves from their interpassive stupor. The informal censorship internalized and propagated by the cultural workers of late capitalism generates a banal conformity that the propaganda chiefs of Stalinism could only ever have dreamt of imposing. Zer0 Books knows that another kind of discourse – intellectual without being academic, popular without being populist – is not only possible: it is already flourishing, in the regions beyond the striplit malls of so-called mass media and the neurotically bureaucratic halls of the academy. Zer0 is committed to the idea of publishing as a making public of the intellectual. It is convinced that in the unthinking, blandly consensual culture in which we live, critical and engaged theoretical reflection is more important than ever before.